W9-AHS-580

THE GLOBAL POSITIONING SYSTEM
A Shared National Asset

Recommendations for Technical Improvements and Enhancements

Committee on the Future of the Global Positioning System

Commission on Engineering and Technical Systems

National Research Council

NATIONAL ACADEMY PRESS

Washington, D.C. 1995

NOTICE: The project that is the subject of this report was approved by the Governing Board of the National Research Council, whose members are drawn from the councils of the National Academy of Sciences, the National Academy of Engineering, and the Institute of Medicine. The members of the committee responsible for the report were chosen for their special competencies and with regard for appropriate balance.

This report has been reviewed by a group other than the authors according to procedures approved by a Report Review Committee consisting of members of the National Academy of Sciences, the National Academy of Engineering, and the Institute of Medicine.

The National Academy of Sciences is a private, nonprofit, self-perpetuating society of distinguished scholars engaged in scientific and engineering research, dedicated to the furtherance of science and technology and to their use for the general welfare. Upon the authority of the charter granted to it by the Congress in 1863, the Academy has a mandate that requires it to advise the federal government on scientific and technical matters. Dr. Bruce M. Alberts is president of the National Academy of Sciences.

The National Academy of Engineering was established in 1964, under the charter of the National Academy of Sciences, as a parallel organization of outstanding engineers. It is autonomous in its administration and in the selection of its members, sharing with the National Academy of Sciences the responsibility for advising the federal government. The National Academy of Engineering also sponsors engineering programs aimed at meeting national needs, encourages education and research, and recognizes the superior achievements of engineers. Dr. Robert M. White is president of the National Academy of Engineering.

The Institute of Medicine was established in 1970 by the National Academy of Sciences to secure the services of eminent members of appropriate professions in the examination of policy matters pertaining to the health of the public. The Institute acts under the responsibility given to the National Academy of Sciences by its congressional charter to be an adviser to the federal government and, upon its own initiative, to identify issues of medical care, research, and education. Dr. Kenneth I. Shine is president of the Institute of Medicine.

The National Research Council was organized by the National Academy of Sciences in 1916 to associate the broad community of science and technology with the Academy's purposes of furthering knowledge and advising the federal government. Functioning in accordance with general policies determined by the Academy, the Council has become the principal operating agency of both the National Academy of Sciences and the National Academy of Engineering in providing services to the government, the public, and the scientific and engineering communities. The Council is administered jointly by both Academies and the Institute of Medicine. Dr. Bruce M. Alberts and Dr. Robert M. White are chairman and vice-chairman, respectively, of the National Research Council.

Library of Congress Catalog Card Number 95-69627
International Standard Book Number: 0-309-05283-1

Available in limited supply from:
The Aeronautics and Space Engineering Board
2101 Constitution Avenue, N.W.
Washington, D.C. 20418
(202) 334-2855

Additional copies available for sale from:
National Academy Press
2101 Constitution Avenue, N.W., Box 285
Washington, D.C. 20055
1-800-624-6242 or (202) 334-3313

Copyright 1995 by the National Academy of Sciences. All rights reserved.
Printed in the United States of America

Copies of *The Global Positioning System — Charting the Future* are available for sale from:
National Academy of Public Administration
Publications Desk
1120 G Street, NW
Suite 850
Washington, D.C. 20005-3801
(202) 347-3190

COMMITTEE ON THE FUTURE OF THE GLOBAL POSITIONING SYSTEM

Laurence J. Adams, *Chair*, Martin Marietta Corporation (Ret.), Consultant, Potomac, Maryland
Penina Axelrad, Colorado Center for Astrodynamics Research, University of Colorado, Boulder, Colorado
John D. Bossler, Center for Mapping, Ohio State University, Columbus, Ohio
Ronald Braff, Center for Advanced Aviation, System Development, MITRE Corporation, McLean, Virginia
A. Ray Chamberlain, American Trucking Association, Inc., Alexandria, Virginia
Ruth M. Davis, Pymatuning Group, Inc., Alexandria, Virginia
John V. Evans, COMSAT Laboratories, COMSAT Corporation, Clarksburg, Maryland
John S. Foster, TRW Inc. (Retired), Redondo Beach, California
Emanuel J. Fthenakis, Fairchild Industries (Ret.), Potomac, Maryland
J. Freeman Gilbert, Institute of Geophysics and Planetary Physics, University of California, San Diego, La Jolla, California
Ralph H. Jacobson, The Charles Stark Draper Laboratory, Inc., Cambridge, Massachusetts
Keith D. McDonald, Sat Tech Systems, Arlington, Virginia
Irene C. Peden, University of Washington, (Retired) Seattle, Washington
James W. Sennott, Department of Electrical and Computer Engineering and Technology, Bradley University, Peoria, Illinois
Joseph W. Spalding, U.S. Coast Guard Research and Development Center, Groton, Connecticut
Lawrence E. Young, Jet Propulsion Laboratory, Pasadena, California

Staff

Archie Wood, Executive Director, Commission on Engineering and Technical Systems
JoAnn C. Clayton, Director, Aeronautics and Space Engineering Board
Allison C. Sandlin, Study Director
David A. Turner, Study Consultant
Cristellyn Banks, Project Assistant

COMMISSION ON ENGINEERING AND TECHNICAL SYSTEMS

Albert R. C. Westwood, Research and Exploratory Technology, Sandia National
  Laboratories, Albuquerque, New Mexico, *Chair*
Naomi F. Collins, NAFSA: Association of International Educators, Washington D.C.
Nancy R. Connery, Woolwich, Maine
Richard A. Conway, Union Carbide Corporation, South Charleston, West Virginia
Samuel C. Florman, Kreisler Borg Florman Construction Company, Scarsdale New York
Trevor O. Jones, Libbey-Owens-Ford Company, Cleveland, Ohio
Nancy G. Leveson, Department of Computer Science and Engineering, University of
  Washington, Seattle, Washington
Alton D. Slay, Slay Enterprises, Inc., Warrenton, Virginia
James J. Solberg, Purdue University, West Lafayette, Indiana
Barry M. Trost, Chemistry Department, Stanford University, Stanford, California
George L. Turin, Berkeley, California
William C. Webster, College of Engineering, Berkeley, California
Deborah A. Whitehurst, Arizona Community Foundation, Phoenix Arizona
Robert V. Whitman, Lexington, Massachusetts

Staff

Archie Wood, Executive Director, Commission on Engineering and Technical Systems

Acknowledgements

The National Research Council's Committee on the Future of the Global Positioning System would like to thank all the individuals who participated in this study, especially Mr. Jules McNeff, Major Lee Carrick, Major Matthew Brennen, Lieutenant Brian Knitt, Captain Earl Pilloud, Captain Christopher Shank, Lieutenant Colonel Donald Latterman, Major Al Mason, Mr. John Clark, Mr. Scott Feairheller, Mr. Terry McGurn, Mr. Jim Graf, Mr. John Hrinkevich, and Mr. Jon Schnabel who arranged briefings and responded to committee requests throughout the study. In addition, Mr. Peter Serini and Mr. George Wiggers served as the committee's liaisons with the Department of Transportation and also were helpful in obtaining relevant information and arranging briefings. The NRC committee also benefited from the work of numerous previous study groups, and considered their recommendations. In addition to the many informative briefings, the committee requested a large number of written responses from receiver manufacturers and many others concerning various issues. The NRC committee wishes to thank all of the contributors for their cooperation in providing existing information and in researching some of the issues that arose. The committee also would like to acknowledge Mr. Michael Dyment of Booz•Allen & Hamilton, who conducted an analysis of the economic impact of the removal of Selective Availability on the differential GPS market; Mr. Melvin Barmat of Jansky/Barmat Telecommunications, Inc., who performed an analysis of L-band frequency availability; and Dr. Young Lee of the MITRE Corporation, who conducted an analysis of the effect of improved accuracy on Receiver Autonomous Integrity Monitoring. A complete list of study participants is given in Appendix A.

Preface

The Global Positioning System (GPS) was originally designed primarily to provide highly accurate radionavigation capability to U.S. military forces, while also providing an unencrypted signal of degraded accuracy to civilian users. As the system developed, civil usage expanded rapidly and the number of civilian users now greatly exceeds the number of military users. The timing, velocity, and positioning information provided by GPS is being used for a growing number of new, innovative applications that could not have been foreseen by the original system designers. Because of its widespread use by both the military and civilians, GPS has truly emerged as a dual-use system.

Recognizing that the continued existence of GPS as a dual-use system clearly requires some trade-offs between civilian utility and national security, Congress requested a joint study by the National Academy of Sciences and the National Academy of Public Administration (NAPA) on the Department of Defense's Global Positioning System (GPS). The National Academy of Sciences was asked to recommend technical improvements and augmentations that could enhance military, civilian, and commercial use of the system. The National Academy of Public Administration was asked to address GPS management and funding issues, including commercialization, governance, and international participation. To conduct its part of the study, the National Academy of Sciences established an expert committee through the National Research Council (NRC), the operating arm of the National Academy of Sciences and the National Academy of Engineering.

This report provides the results of the technical portion of the study conducted by the National Research Council's Committee on the Future of the Global Positioning System. Portions of this report (for example, Chapters 3, 4, and some of the appendices) also are included in the joint NRC/NAPA report, *The Global Positioning System — Charting the Future*, which contains the complete results of the NAPA portion of the study.

In examining future enhancements to the GPS system, the NRC committee endeavored to balance the features that would enhance civil applications against the clear requirement to maintain the military integrity of the system. The recommendations in the report were intended to meet this criterion.

Laurence J. Adams, Chair
Committee on the Future of
the Global Positioning System

Table of Contents

List of Figures

List of Tables

Acronyms and Abbreviations

ADS   Automatic dependant surveillance
ANSI   American National Standards Institute
A-S   Anti-Spoofing
ASIC   Application Specific Integrated Circuit
ATM   Air Traffic Management
AVI   Automatic Vehicle Identification
AVL   Automatic Vehicle Location
BIPM   *Bureau International des Poids et Measures*
C/A   Coarse/Acquisition code
CDMA   Code Division Multiple Access
CEP   Circular Error Probable
CGS   Civil GPS Service
CGSIC   Civil GPS Service Interface Committee
CORS   Continuously Operating Reference Station
CRPA   Controlled Radiation (Reception) Patterned Antenna
dB   decibel
DGPS   Differential GPS
DMA   Defense Mapping Agency
DOD   Department of Defense
DOP   Dilution of Precision
DOT   Department of Transportation
drms   distance root mean square
DRVID   Differential Ranging Versus Integrated Doppler
ECDIS   Electronic Chart Display Information System
FAA   Federal Aviation Administration (part of DOT)
FDMA   Frequency Division Multiple Access
FHWA   Federal Highway Administration
FM   Frequency Modulation
FRA   Federal Railroad Administration
GIS   Geographic Information Systems
GHz   Gigahertz
GLONASS   GLObal Navigation Satellite System

GNSS	Global Navigation Satellite System
GPS	Global Positioning System
HDOP	Horizontal Dilution of Precision
Hz	Hertz (cycles per second)
IALA	International Association of Lighthouse Authorities
ICAO	International Civil Aviation Organization
IF	Intermediate Frequency
IGS	International GPS Service for Geodynamics
ILS	Instrument Landing System
IMO	International Maritime Organization
Inmarsat	International Maritime Satellite Organization
INS	Inertial Navigation System
ITS	Intelligent Transportation System
IVHS	Intelligent Vehicle Highway Systems
JCS	Joint Chiefs of Staff
JPO	Joint Program Office
J/S	Jammer-to-signal ratio
KHz	Kilohertz
L_1	GPS L-band signal 1 (1575.42 Mhz)
L_2	GPS L-band signal 2 (1227.6 MHz)
L_4	Proposed GPS L-band signal
L-band	L-band frequency (about 1-2 GHz)
LADGPS	Local Area Differential GPS
LORAN-C	Long-Range Navigation, Version C
MBS	Mobile Broadcast Service
MCS	GPS Master Control Station
MHz	Megahertz
ms	Millisecond
MOA	Memorandum of Agreement
NAPA	National Academy of Public Administration
NASA	National Aeronautics and Space Administration
NCA	National Command Authority
NDB	Nondirectional Beacon
NGS	National Geodetic Survey
NIST	National Institute of Standards and Technology
NOAA	National Oceanic and Atmospheric Administration
NRC	National Research Council
ns	nanosecond
NSA	National Security Agency
NTIA	National Telecommunications and Information Administration
OCS	Operational Control Segment
P-code	Precision code
P_{HE}	Probability of Hazardous Error
PLGR	Precision Lightweight GPS Receiver

P_{MD}	Probability of Missed Detection
P^3I	Preplanned Product Improvement
PPS	Precise Positioning Service
PRN	Pseudorandom Noise
RAIM	Receiver Autonomous Integrity Monitoring
RDS	Radio Data System
RF	Radio Frequency
RFP	Request for Proposal
RISC	Reduced Instruction Set Computing
RNP	Required Navigation Performance
ROD	Relative Operating Distance
RTCA	Ratio Technical Commission for Aeronautics
RTCM	Radio Technical Commission for Maritime Services
SA	Selective Availability
SAIM	Satellite Autonomous Integrity Monitoring
S-band	Microwave frequency band, about 2-4 Ghz
SEP	Spherical Error Probable
sigma	standard deviation (symbol: σ)
SNR	Signal-to-Noise Ratio
SONET	Synchronized Optical Network
SPS	Standard Positioning Service
TACAN	Tactical Air Navigation
TAI	International Atomic Time
TCAS	Traffic Alert/Collision Avoidance System
TDMA	Time Division Multiple Access
TEC	Total Electron Content
TOD	Time of day
UERE	User Equivelent Range Error
UHF	Ultra High Frequency
USAF	United States Air Force
USCG	United States Coast Guard
USNO	United States Naval Observatory
UTC	Coordinated Universal Time
VDOP	Vertical Dilution of Precision
VHF	Very High Frequency
VLBI	Very Long Baseline Interferometry
VOR	VHF Omnidirectional Range
VOR/DME	VOR with Distance Measuring Equipment
VORTAC	combined VOR and TACAN
VTS	Vessel Traffic Services
WAAS	Wide Area Augmentation System
WADGPS	Wide Area Differential GPS
WGS	World Geodetic System
Y-code	Encrypted P-code

Executive Summary

In response to a request from Congress, a joint study on the Department of Defense's Global Positioning System (GPS) was conducted by the National Academy of Sciences and the National Academy of Public Administration. The National Academy of Sciences was asked to recommend technical improvements and augmentations that could enhance military, civilian, and commercial use of the system. The National Academy of Public Administration was asked to address GPS management and funding issues, including commercialization, governance, and international participation. To conduct its part of the study, the National Academy of Sciences established an expert committee, through the National Research Council (NRC), the operating arm of the National Academy of Sciences and the National Academy of Engineering.

Specifically, the National Academy of Sciences was asked to address the following three technical questions:

(1)   Based on presentations by the Department of Defense (DOD) and the intelligence community on threats, countermeasures, and safeguards, what are the implications of such security-related safeguards and countermeasures for the various classes of civilian GPS users and for future management of GPS? In addition, are the Selective Availability and Anti-Spoofing capabilities of the GPS system meeting their intended purpose?

(2)   What augmentations and technical improvements to the GPS itself are feasible and could enhance military, civilian, and commercial use of the system?

(3)   In order to preserve and promote U.S. industry leadership in this field, how can communication, navigation, and computing technology be integrated to support and enhance the utility of GPS in all transportation sectors, in scientific and engineering applications beyond transportation, and in other civilian applications identified by the study in the context of national security considerations?

In its interpretation of Task 1, the NRC committee decided not only to determine whether Selective Availability (SA) and Anti-Spoofing (A-S) were meeting their intended purpose, but also to determine the broad ramifications of the use of these techniques and to make specific recommendations for each. In response to Task 2, the committee made recommendations for technical improvements because it believed that only identification of technical improvements would be of little value without an accompanying recommendation. In response to Task 3, the NRC committee considered "U.S. industry leadership" to mean technical preeminence focused on meeting the demands of a growing number of user applications, while maintaining a technical advantage for the DOD.

TASK 1

Based on presentations by the DOD and the intelligence community on threats, countermeasures, and safeguards, what are the implications of such security-related safeguards and countermeasures for the various classes of civilian GPS users and for future management of GPS? In addition, are the Selective Availability and Anti-Spoofing capabilities of the GPS system meeting their intended purpose?

The DOD has stated that SA[1] is an important security feature because it prevents a potential enemy from directly obtaining positioning and navigation accuracy of 30 meters (95 percent probability) or better from the C/A-code.[2] Since the military has access to a specified accuracy of 21 meters (95 percent probability), they believe U.S. forces have a distinct strategic and tactical advantage. With SA at its current level, a potential enemy has access only to the C/A-code signal with a degraded accuracy of only 100 meters (95 percent probability). The DOD believes that obtaining accuracies better than 100 meters (95 percent probability) requires a substantial amount of effort on the part of an unauthorized user. Further, DOD representatives have expressed their belief that our adversaries are much more likely to exploit the GPS C/A-code rather than differential GPS (DGPS), because its use requires less effort and technical sophistication than is required to use DGPS.[3] In addition, some DOD representatives contend that local-area DGPS broadcasts do not

---

[1] SA is a purposeful degradation in GPS navigation and timing accuracy that is accomplished by intentionally varying the precise time of the clocks on board the satellites, which introduces errors into the GPS signal. With SA, the civilian signal on which the Coarse Acquisition (C/A) code is transmitted, is limited to an accuracy of 100 meters, 95 percent probability. Military receivers with the appropriate encryption keys can eliminate the effects of SA and obtain an accuracy of approximately 21 meters (95 percent probability).

[2] The Coarse Acquisition (C/A) code is broadcast on the L-band carrier signal known as L_1, which is centered at 1575.42 MHz.

[3] DGPS is based upon knowledge of the highly accurate, geodetically surveyed location of a GPS reference station, which observes GPS signals in real time and compares their ranging information to the ranges expected to be observed at its fixed point. The differences between observed ranges and predicted ranges are used to compute corrections to GPS parameters, error sources, and/or resultant positions. These differential corrections are then transmitted to GPS users, who apply the corrections to their received GPS signals or computed position.

diminish the military advantage of SA because they could be rendered inoperative, if warranted, through detection and destruction or by jamming.

It is opinion of the NRC committee, however, that any enemy of the United States sophisticated enough to operate GPS-guided weapons will be sophisticated enough to acquire and operate differential systems. Enemies could potentially take advantage either of the existing, commercial systems available worldwide or install a local DGPS system, which could be designed and operated in a manner that would be difficult to detect. These systems can have the capability to provide velocity and position corrections to cruise and ballistic missiles with accuracies that are equal to or superior to those available from an undegraded C/A-code. It should be noted that with both GPS- and DGPS-guided weapons, accurate knowledge of the target location is a prerequisite for weapon accuracy. Even if the level of SA is increased, DGPS methods could still be used to provide an enemy with accurate signals. Thus, the NRC committee concluded that the existence and widespread proliferation of DGPS augmentations have significantly undermined the effectiveness of SA in denying accurate radionavigation signals to our adversaries. In addition, the Russian GLONASS system broadcasts unencrypted signals with an accuracy comparable to an undegraded GPS C/A-code, which further erodes the effectiveness of SA.[4]

The unencrypted C/A-code, which is degraded by SA, still provides our adversaries with an accuracy of 100 meters (95 percent probability). With SA set at zero, the stand-alone accuracy improves to around 30 meters (95 percent probability).[5] While this improvement enhances the ability of an adversary to successfully attack high-value point targets, significant damage also can be inflicted with accuracies of 100 meters, (95 percent probability). Therefore, in either case (30-meter or 100-meter accuracy) the risk is sufficiently high to justify denial of the L_1 signal by jamming. The jamming strategy has the additional benefit of denying an adversary all radionavigation capability, including the even more accurate DGPS threat.

The NRC committee strongly believes that preservation of our military advantage with regard to radionavigation systems should focus on electronic *denial* of all useful signals to an opponent, for example, by jamming and spoofing, while improving the ability of civil and friendly military users to employ GPS in a jamming and spoofing environment. Continued effort to deny the accuracy of GPS to all users except the U.S. military via SA appears to be a strategy that ultimately will fail. Thus, the NRC committee recommends that the military employ denial techniques in a theater of conflict to prevent enemy use of GPS or other radionavigation systems.

---

[4] GLObal Navigation Satellite System or GLONASS is a space-based radionavigation system also consisting of three segments just as GPS does. GLONASS is operated and managed by the military of the former Soviet Union. The GLONASS space segment also is designed to consist of 24 satellites arranged in three orbital planes. The full GLONASS constellation is currently scheduled to be completed in 1995. GLONASS does not degrade the accuracy of its civilian signal by SA or similar techniques.

[5] Recent measurements with SA turned to zero have ranged from 5 meters to 10 meters (95 percent probability). However, the accuracy without SA greatly depends on the condition of the ionosphere at the time of observation and user equipment capabilities.

The NRC committee believes that the principal shortcoming in a denial strategy, regardless of the level of SA, is the difficulty that military GPS receivers currently have in acquiring the Y-code during periods when the C/A-code is unavailable due to jamming of the L_1 signal. The implementation of direct Y-code acquisition capability, as recommended in Chapter 3, would provide the optimal solution to this problem. The technology for developing direct Y-code receivers is available today. The committee believes that a focused, high-priority effort by the DOD to develop and deploy direct Y-code user equipment, backed by forceful political will from both the legislative and executive branches, can bring about the desired result in a relatively short period of time. In the interim before direct Y-code receivers can be fielded by the military, various operating disciplines, also discussed in Chapter 3, can be used to minimize the impact of L_1 C/A-code jamming on the ability to acquire the Y-code directly.

From the onset of the study, the NRC committee agreed that national security was of paramount importance and, without exception, the U.S. military advantage should be maintained. As outlined above, the committee determined that the military effectiveness of SA is greatly diminished because of the widespread proliferation of DGPS and existence of GLONASS. In addition, the NRC committee compiled the following findings related to the effects of SA and A-S[6] on the various classes of civilian users:

- The presence of SA and A-S increases the cost and complexity of Federal Aviation Administration's Wide Area Augmentation System (WAAS)[7] and limits the effectiveness of Receiver Autonomous Integrity Monitoring (RAIM).[8]

- The presence of SA affects the acceptance of GPS by some commercial users and limits the ability of the Coast Guard's DGPS service to provide important safety-related information to its users.

- GPS-based automobile navigation systems, which require accuracies in the 5- to 20-meter range, would no longer require DGPS if SA was

---

[6] Anti-Spoofing (A-S) is the encryption process used to deny unauthorized access to the military Y-code. It also significantly improves a receiver's ability to resist locking onto mimicked GPS signals, which could potentially provide incorrect positioning information to a GPS user.

[7] Wide-Area Augmentation System (WAAS) is a wide-area DGPS concept planned by the FAA to improve the accuracy, integrity, and availability of GPS to levels that support flight operations in the National Airspace System from en route navigation through Category I precision approaches. WAAS will consist of a ground-based communications network and several geosynchronous satellites to provide nationwide coverage. The ground-based communications network will consist of 24 wide-area reference stations, two wide-area master stations, and two satellite uplink sites. Differential corrections and integrity data derived from the ground-based network, as well as additional ranging data, will be broadcast to users from the geostationary satellites using an "L_1-like" signal.

[8] Receiver Autonomous Integrity Monitoring (RAIM) is a method to enhance the integrity of a GPS receiver without requiring any external augmentations. RAIM algorithms rely on redundant GPS satellite measurements as a means of detecting unreliable satellites or position solutions.

eliminated and further improvements were made to the basic GPS as suggested in Chapter 3. The elimination of SA would also improve the performance of those DGPS systems required for higher-accuracy applications, such as collision avoidance, that are important to the future Intelligent Transportation System.

- Most mapping, surveying, and geodetic applications would be enhanced by cost savings from quicker acquisition of data. The elimination of SA and the ability to track code on two frequencies can improve acquisition time.

- Post-processing can eliminate the effects of SA for most Earth science applications, but the presence of A-S increases the cost and limits the performance of many techniques.

- Although GPS currently meets all accuracy requirements for both GPS time transfer and time synchronization using direct GPS time, many telecommunications companies are still hesitant to utilize GPS because of concerns about system reliability and the presence of SA.

- SA has little or no effect on the ability to use GPS for spacecraft orbit or attitude determination, but A-S limits the performance of orbit determination for spacecraft that rely on dual-frequency codeless measurements. A-S may also contribute to limitations on achievable attitude determination accuracy.

The six most important findings of the NRC committee regarding the impact of SA on the various classes of civilian users and on meeting its intended purpose are

(1) The military effectiveness of SA is significantly undermined by the existence and widespread proliferation of DGPS augmentations as well as the potential availability of GLONASS signals.

(2) Turning SA to zero would have an immediate positive impact on civil GPS users. Without SA, the use of DGPS would no longer be necessary for many applications. System modifications that would further improve civilian accuracy also would be possible without SA.

(3) Deactivation of SA would likely be viewed as a good faith gesture by the civil community and could substantially improve international acceptance and potentially forestall the development of rival satellite navigation systems.

Without SA, the committee believes that the number of GPS and DGPS users in North America would increase substantially.[9]

(4)   It is the opinion of the committee that the military should be able to develop doctrine, establish procedures, and train troops to operate in an L_1 jamming environment in less than three years.

(5)   The technology for developing direct Y-code receivers is currently available and the development and initial deployment of these receivers could be accomplished in a short period of time if adequately funded.

(6)   The FAA's WAAS, the Coast Guard's differential system, and GLONASS are expected to be fully operational in the next 1 to 3 years. The Coast Guard's DGPS network and the WAAS will provide accuracies greater than that available from GPS with SA turned to zero and GLONASS provides accuracies that are comparable to GPS without SA. At the same time, other local DGPS capabilities are likely to continue to proliferate.

Selective Availability should be turned to zero immediately and deactivated after three years. In the interim, the prerogative to reintroduce SA at its current level should be retained by the National Command Authority.

Although many civil users could benefit if A-S is turned off as noted above, the NRC committee found that A-S remains critically important to the military because it forces potential adversaries to use the C/A-code on L_1, which can be jammed if necessary without inhibiting the U.S. military's use of the encrypted Y-code on L_2. Further, encryption provides resistance to spoofing of the military code. The NRC committee determined, however, that the current method of manual distribution of Y-code decryption keys is laborious and time consuming. The DOD has recognized this problem and has ongoing efforts to distribute keys electronically. The NRC committee believes that an electronic key distribution capability would greatly enhance the use of the encrypted L_2 Y-code. The committee also believes that technology is available to upgrade the current encryption method and suggests that the Air Force should explore the necessity of utilizing this technology. Modifications to the Block IIR satellites and the Block IIF request for proposal may be required if upgraded encryption methods are necessary. Changes to military receivers also will be required.

A-S should remain on and the electronic distribution of keys should be implemented at the earliest possible date. In addition, the Air Force should explore the necessity of upgrading the current encryption method. Required receiver enhancements should be incorporated in future planned upgrades.

---

[9] The analysis by Michael Dyment, Booz•Allen & Hamilton, 1 May 1995, is shown in Appendix E.

TASK 2

What augmentations and technical improvements to the GPS itself are feasible and could enhance military, civilian, and commercial use of the system?

Today GPS is a true dual-use system. Although it was originally designed to provide a military advantage for U.S. forces, the number of civilian users now exceeds the number of military users. During the course of the study, the NRC committee examined various technologies and augmentations applicable to GPS. It determined that several improvements could be made to the system that would enhance its use for civilian, commercial, and military users without compromising national security. Some of the improvements could be made immediately; others could be incorporated on some of the Block IIR spacecraft that are currently being built and included in the specification requirements for the next generation Block IIF spacecraft. The committee's recommendations are listed below and a detailed discussion of each is provided in Chapter 3. Although the approximate cost of each improvement is given when available, potential funding mechanisms for each improvement are not discussed. In general, the issue of GPS funding is addressed by the National Academy of Public Administration.

Recommendations that Enhance GPS Performance for Civil and Commercial Users

The NRC committee found that the most prominent need for commercial and civil users is greater stand-alone accuracy, availability, and integrity. With improved performance of the basic GPS signal, many users would no longer require augmentations to obtain the data they require. Any additional system enhancements and modifications to improve stand-alone positioning accuracy for civilian users are relatively ineffective in the presence of SA. However, if the recommendation to deactivate SA is implemented, the committee has identified several enhancements that could provide significant improvement for both civilian and military users. With SA removed, the major enhancement that would greatly increase accuracy for civilian users is the addition of a new, unencrypted signal that allows for corrections of errors introduced by the ionosphere.[10] While very important for civil users, this feature will provide minimal additional capability to military users because they already have this capability through use of their encrypted signals.

Immediate steps should be taken to obtain authorization to use an L-band frequency for an additional GPS signal, and the new signal should be added to GPS Block IIR satellites at the earliest opportunity.

---

[10] A preliminary analysis of the L-band spectrum allocation that was conducted by Mr. Melvin Barmat, Jansky/Barmat Telecommunications Inc., Washington D.C., January 1994, is shown in Appendix I.

Recommendations that Enhance GPS Performance for Military Users

As stated above, GPS was originally designed to provide our forces with a military advantage. In the past, DOD has depended on a strategy of global signal degradation, through SA, to reduce the GPS signal accuracy to civilian and unauthorized users, while providing a more accurate, encrypted signal to authorized users. However, as stated above, the committee believes that the military usefulness of SA is severely diminished and that it is urgent that the DOD focus its attention on *denial* of all useful signals to an opponent, for example, through jamming and spoofing techniques, including jamming of the unencrypted C/A-code, rather than relying on SA. The NRC committee therefore recommends several military receiver enhancements that would support such a strategy.

> *The development of receivers that can rapidly lock onto the Y-coded signals in the absence of the C/A-code should be completed. The deployment of direct Y-code receivers should be given high priority by the DOD.*

> *Nulling antennas and antenna electronics should be employed whenever feasible and cost effective. Research and development focused on reducing the size and cost of this hardware should actively be supported.*

> *The development of low-cost, solid-state, tightly-coupled integrated inertial navigation system/GPS receivers to improve immunity to jamming and spoofing should be accelerated.*

> *The development and operational use of GPS receivers with improved integration of signal processing and navigation functions for enhanced performance in jamming and spoofing should be accelerated.*

> *Military receivers should be developed that compensate for ionospheric errors when L_1 is jammed, by improved software modeling and use of local-area ionospheric corrections.*

In the interim time before such enhancements can be fielded by the military, various operating disciplines, which are discussed in Chapter 3, can be used to minimize the impact of C/A-code jamming on the ability to acquire the Y-code directly.

Recommendations that Enhance GPS Performance for All Users (Civil, Commercial, and Military)

In view of the rapidly expanding use of GPS, the NRC committee believes that GPS must be capable of continuous operation in all foreseeable contingencies. This capability is critical. The one area where the NRC committee found limited redundancy was in the operational control segment (OCS). Although the NRC committee determined that the Air

Force has several experiments planned to improve the system, it believes there are some additional improvements that can be made to the OCS that would increase stand-alone accuracy, availability, and integrity; improve the overall reliability of the system; or simplify day-to-day operations. Recommendations that would result in greater stand-alone GPS accuracy and integrity include uploading more current clock and orbit information to all satellites, increasing the number of monitor sites, reducing the clock and ephemeris errors, and improving Block IIR and Block IIF integrity monitoring capability. In addition, the NRC committee found a need for (1) a simulator to test software and train personnel, (2) modern receivers at the monitor stations, and (3) a permanent, backup master control station. Specifically, the NRC committee recommends:

Additional GPS monitoring stations should be added to the existing operational control segment. Comparison studies between cost and location should be completed to determine if Defense Mapping Agency or Air Force sites should be used.

The operational control segment Kalman Filter should be improved to solve for all GPS satellites' clock and ephemeris errors simultaneously through the elimination of partitioning, and the inclusion of more accurate dynamic models. These changes should be implemented in the 1995 OCS upgrade request for proposal.

Procurements for the replacement of the monitor station receivers, computers, and software should be carefully coordinated. The new receivers should be capable of tracking all satellites in view and providing C/A-code, Y-code, and L_1, and L_2 carrier observables to the OCS. Upgradability to track a new L_4 signal also should be considered. OCS software also should be made capable of processing this additional data.

Firm plans should be made to ensure the continuous availability of a backup master control station.

A simulator for the space and ground segment should be provided as soon as possible to test software and train personnel.

The operational control segment software should be updated using modern software engineering methods in order to permit easy and cost-effective updating of the system and to enhance system integrity. This should be specified in the 1995 OCS upgrade request for proposal.

The planned Block IIR operation should be reexamined and compared to the accuracy advantages gained by incorporating inter-satellite ranging data in the ground-based Kalman Filter and uploading data at some optimal time interval, such as every hour, to all GPS satellites.

Block IIR satellite communication crosslinks should be used to the extent possible with the existing crosslink data rate to support on-board satellite health monitoring for improved reliability and availability and in order to permit a more rapid response time by the operational control segment.

The Block IIR inter-satellite communication crosslinks should be used to relay integrity information determined through ground-based monitoring.

The DOD's more frequent satellite navigation correction update strategy should be fully implemented as soon as possible following the successful test demonstration of its effectiveness. In addition, the current security classification policy should be examined to determine the feasibility of relaxing the 48-hour embargo on the clock and ephemeris parameters to civilian users.

If the above recommendations are implemented, the NRC committee believes that the overall GPS performance and reliability will be greatly enhanced and that a stand-alone horizontal accuracy of the basic GPS signal approaching 5 meters (95 percent probability) could be achieved for both civilian and military users.

Improvement Implementation Strategy

Because of the relatively long life time of GPS satellites (5 to 10 years) and the length of time required to replace the total constellation of 24 satellites, opportunities for introducing enhancements and technology improvements to the system are limited.

Figure 1 shows the current plan for satellite replacements. According to the GPS Joint Program Office, current plans for the Block IIF contract include 6 short-term and 45 long-term "sustainment" satellites. As currently planned, the Block IIF satellites will be designed to essentially the same specifications as the Block IIR satellites. The current program and schedule make it possible for another country to put up a technically superior system that uses currently available technology before the United States can do so. Under the current planning and in the absence of a preplanned product improvement (P^3I) program, the earliest opportunity for an infusion of new technology in the GPS space segment would be after Block IIF, probably sometime after the year 2020.

The NRC committee believes that there are significant improvements that could be made to the system much earlier than post-Block IIF that would not only enhance its performance for civilian and military use but also make it more acceptable and competitive internationally. One method to incorporate technology in an efficient and timely manner is through a P^3I program beginning as early as possible in Block IIR. With this type of approach, planned changes and improvements could be intentionally designed into the production of the satellites at specific time intervals.

Assuming that the first improvements suggested in this report are incorporated in the later half of the Block IIR satellites, additional funding might be required to incorporate changes for the already completed Block IIR satellites. However, the NRC committee believes that the timely improvement in system performance is adequate justification for the additional cost.

In addition to the specific recommendations given in this report, the NRC committee also discussed several enhancements that it believes have particular merit and should be seriously considered for future incorporation. These items are discussed in Chapter 4. Although a few enhancements could be included on the Block IIR spacecraft, especially if a P^{3}I program were implemented, most of the enhancements would have to be incorporated in the Block IIF spacecraft design.

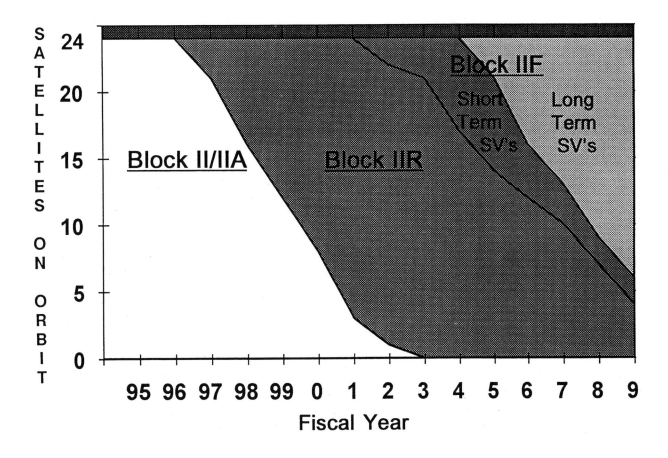

Figure 1 Current Plan for Satellite Replacement. (Courtesy of the GPS Joint Program Office)

TASK 3

In order to preserve and promote U.S. industry leadership in this field, how can communication, navigation, and computing technology be integrated to support and enhance the utility of GPS in all transportation sectors, in scientific and engineering applications beyond transportation, and in other civilian applications identified by the study in the context of national security considerations?

As described in Chapter 2 and Appendix C, the NRC committee found that civil, commercial, and military GPS users are making rapid progress in developing and utilizing systems that integrate GPS with other technologies. For many navigation and position location applications, GPS is being combined with one or more of the following: radar; inertial navigation systems; dead reckoning systems; aircraft avionics and flight management systems; digital maps; computers and computer databases; and communication datalinks. For timing applications, GPS can be combined with reference clocks and digital communication networks. Surveying and mapping users have combined GPS with computer databases, inertial navigation systems, digital imaging systems, and laser measuring systems. Earth science users have integrated GPS with radar altimeters, precision accelerometers, synthetic aperture radar, computer databases and workstations, and communications datalinks.

By integrating GPS with other technologies, highly accurate positioning and timing information can be obtained at a very modest cost, which provides a large incentive to system designers to develop integrated GPS products. For example, with the large market potential for ground vehicle position location and guidance systems, there is considerable motivation for the vigorous commercially funded research and development activity that is underway. The NRC committee believes that the U.S. user equipment industry's intensive focus on research and development is sufficient to ensure that its technical competitiveness will be maintained.

During its deliberations, the committee found that some user communities had a limited number of very specific issues related to the integrated use of GPS with other technologies that may require government action. Examples include the need to modernize the air traffic management system to take advantage of the full capabilities of GPS-based navigation and surveillance and the need to speed up the process of providing up-to-date digital hydrographic data for use in Electronic Chart Display Information Systems (ECDIS). These findings and others have been reported in Chapter 2. In general, however, the GPS industry is meeting most user demands by continuously improving integrated user equipment and services and is limited only by the need to augment and enhance the characteristics of the basic GPS constellation. Therefore, it is the opinion of the NRC committee that the *most important* government action required is to improve the performance of the basic GPS satellite system to provide the highest levels of position accuracy, signal integrity, and signal availability that can be technologically achieved at reasonable cost without negatively impacting national security. The committee believes that the performance improvements summarized in response to Task 2 above and further discussed in Chapter 3 meet these criteria.

1

Introduction

THE TASK

In the National Defense Authorization Act for fiscal year 1994, the congressional committees that authorize the activities of the Department of Defense (DOD) requested that a joint study on the Global Positioning System (GPS) be conducted by the National Academy of Sciences and the National Academy of Public Administration (NAPA).[1] The National Academy of Sciences was asked to recommend technical improvements and augmentations that could enhance military, civilian, and commercial use of the system. NAPA was asked to address GPS management and funding issues, including commercialization, governance, and international participation.

Specifically, the National Academy of Sciences was asked to address the following three technical questions:

(1) Based on presentations by the DOD and the intelligence community on threats, countermeasures, and safeguards, what are the implications of such security-related safeguards and countermeasures for the various classes of civilian GPS users and for future management of GPS? In addition, are the Selective Availability and Anti-Spoofing capabilities of the GPS system meeting their intended purpose?

(2) What augmentations and technical improvements to the GPS itself are feasible and could enhance military, civilian, and commercial use of the system?

(3) In order to preserve and promote U.S. industry leadership in this field, how can communication, navigation, and computing technology be integrated to support and enhance the utility of GPS in all transportation sectors, in

---

[1] The National Academy of Sciences carries out its studies through the National Research Council (NRC), the operating arm of the National Academy of Sciences and the National Academy of Engineering, using a committee of experts and a small staff. NAPA utilizes its own staff members and consultants to conduct its studies, which are reviewed throughout the process by an oversight panel of distinguished individuals.

scientific and engineering applications beyond transportation, and in other civilian applications identified by the study in the context of national security considerations?

NAPA was asked to address the following four questions related to future GPS management and funding:

(1)   How should the GPS program be structured and managed to maximize its dual utility for civilian and military purposes?

(2)   How should the GPS program/infrastructure be funded to assure consistent, sustainable, and reliable services to civilian and military users around the world? In consideration of its worldwide user community, are there equitable cost-recovery mechanisms that may be implemented to make the GPS program partially or fully self-supporting without compromising U.S. security or international competitive interests?

(3)   Is commercialization or privatization of all or parts of the GPS consistent with U.S. security, safety, and economic interests?

(4)   Is international participation in the management, operation, and financing of GPS consistent with U.S. security and economic interests?

JOINT STUDY APPROACH

Both the National Academy of Sciences and NAPA are chartered by Congress and conduct studies for the government on issues of national and international importance. National Academy of Sciences studies, which are carried out under the auspices of the National Research Council (NRC), are generally focused on scientific and engineering issues, and NAPA studies are generally focused on management issues. Because Congress was interested in a GPS study that covered both technical and management issues, a joint study was requested in the 1994 Defense Authorization Act.

Because each academy operates differently, the NRC and NAPA portions of the study followed different schedules with different report-writing procedures. Nevertheless, the NRC and NAPA staffs worked closely together throughout the study, drafting joint outlines, exchanging information, attending both the NRC committee and NAPA panel meetings, and meeting frequently to work out details of the joint report. The NRC's peer-review process applied only to the portions of the joint report authored by the NRC committee.

The NRC technical portion of the study began in June 1994 and entered peer review in February 1995. The NAPA management and funding portion began in August 1994 and entered review in March 1995. In May 1995, both reports were combined to form a single document, and the final joint report was delivered to Congress in May 1995.

National Research Council Study Approach

In mid-1994, the NRC formed the Committee on the Future of the Global Positioning System, hereafter referred to as the NRC committee, under the auspices of the Commission on Engineering and Technical Systems. (A membership list is included in the front of this report.) The NRC committee met June 23 through June 25, 1994; July 28 through July 30; August 16 through August 18; September 29 through October 1; October 19 through October 21; November 18; December 15; January 13, 1995; February 11; April 8; and April 17. During these meetings the committee heard over 70 briefings from government officials, industry representatives, commercial interest groups, and technical experts on GPS issues. A complete list of participants is given in Appendix A. Appendix B contains brief biographies for the committee members. In addition, several committee members visited on-site locations to gather additional information and further clarify important issues.

At the first meeting, the NRC committee heard presentations from U.S. Air Force and U.S. Department of Transportation (DOT) representatives. The committee also familiarized itself with the history, management, operation, and technical components of GPS. The second committee meeting focused on the technical requirements of the various civilian and military users. Presentations were made by representatives in the aviation, maritime, transportation, agriculture, surveying and mapping, and scientific communities. Information also was provided by representatives of the precise timing and telecommunications communities. At the third meeting, the committee considered the GPS requirements of the U.S. military services and heard detailed presentations on the GPS space segment, ground control segment, and the user equipment segment. Presentations related to the Russian Global Orbiting Navigation Satellite System (GLONASS) and the U.S. GPS industry also were made at this meeting. The fourth meeting then focused on (1) the threats and vulnerabilities both to and from the use of GPS and (2) GPS jamming, spoofing, and interference issues.

In October and November 1994, and again in January, February, and April 1995, the committee met to discuss its findings and recommendations. In addition, several committee members met on numerous occasions to work on the draft report. On April 17, the NRC committee and the NAPA panel held a joint meeting to finalize the combined report.

MAJOR ISSUES AND CONSIDERATIONS

Although the number of civilian users now exceeds the military users, GPS is a dual-use system that was originally designed to provide our forces with a military advantage. From the onset of this study, the NRC committee agreed that national security was the most critical issue in considering any recommendation in this report and that, without exception, the U.S. military advantage should be maintained. During the course of the study, the NRC committee examined various technologies and augmentations applicable to GPS. The NRC committee determined that several improvements could be made to the system that would enhance its use for civilian, commercial, and military users without compromising national security. Some of the improvements could be made immediately, and others could be incorporated on some of the Block IIR spacecraft that are currently being built and included in the specification requirements for the next generation Block IIF spacecraft.

REPORT ORGANIZATION

In making recommendations, the NRC committee considered the requirements of various civilian and military users. These requirements are discussed in detail in Chapter 2. Recommendations that enhance the basic GPS for all users and recommendations that enhance the basic GPS for specific user groups are presented in Chapter 3. Chapter 4 examines possible enhancements that have particular merit for future incorporation, but which require further study. Data and analyses supporting the NRC committee's recommendations and a more detailed technical overview of GPS and its augmentations and enhancements are compiled in the appendices. Specifically, Appendix C provides a detailed technical and programmatic overview of GPS.

GPS PROGRAM OVERVIEW

The incomparable navigation, positioning, and timing system that is known today as the GPS, is a combination of several satellite navigation systems and concepts developed by or for the U.S. DOD. In 1973, the best characteristics of each of these programs were combined under the auspices of a Joint Program Office (JPO) located at the U.S. Air Force Space and Missile Organization in El Segundo, California. From its inception, the NAVSTAR Global Positioning System was designed to meet the radionavigation requirements of all the military services, and those of civilian users as well.[2]

Responsibility for the day-to-day management of the GPS program and operation of the system continues to rest with the DOD and is carried out primarily by the Air Force.[3] DOD policy for the GPS program is set by the Under Secretary of Defense for Acquisition and Technology, with the help of the DOD Positioning/Navigation Executive Committee. This committee receives input from all the DOD commands, departments, and agencies and coordinates with the DOT Positioning/Navigation Executive Committee and the Assistant Secretary for Transportation Policy. Together, the two secretaries mentioned above make-up the GPS Executive Board.

---

[2] The GPS system is officially known as the NAVSTAR Global Positioning System; however, the NAVSTAR name is rarely used. For the remainder of this report the system will simply be referred to as GPS.

[3] As with all other federally funded navigation systems, the ultimate decision-making authority over GPS operations, in peacetime and in wartime, is the National Command Authority (NCA), consisting of the President or the Secretary of Defense with the approval of the President.

The official source of planning and policy information for each radionavigation service provided by the U.S. government, including GPS, is the *Federal Radionavigation Plan*.[4] The plan is jointly developed by the DOD and the DOT, and is updated biennially. The *Federal Radionavigation Plan* represents an attempt to provide users with the optimal mix of federally provided radionavigation systems and reflects both the DOD's responsibility for national security and the DOT's responsibility for public safety and transportation economy.

GPS TECHNICAL OVERVIEW

The GPS constellation comprises 24 Earth-orbiting satellites, which transmit radio signals that consist of the satellite's position and the time it transmitted the signal. These signals can be received on Earth with a relatively inexpensive device that costs around $400 or so. The distance between a satellite and a receiver can be computed by subtracting the time that the signal left the satellite from the time that it arrives at the receiver. If the distance to four or more satellites is measured, then a three-dimensional position on Earth can be determined. GPS positioning capability is provided at no cost to civilian and commercial users worldwide at an accuracy level of 100 meters (2 drms).[5] This accuracy level is known as the standard positioning service (SPS). The U.S. military and its allies, and a select number of other authorized users, receive a specified accuracy level of 16 meters (SEP), known as the precise positioning service (PPS).[6]

The full accuracy capability of GPS is denied to users of the SPS through a process known as Selective Availability, or SA. SA is the purposeful degradation in GPS navigation accuracy that is accomplished by intentionally varying the precise time of the clocks on board the satellites, which introduces errors into the GPS signal, and by providing incorrect orbital positioning data in the GPS navigation message. SA is normally set to a level that will provide 100-meter (2 drms) positioning accuracy to users of the SPS, as defined in the *Federal Radionavigation Plan*. PPS receivers with the appropriate encryption keys can eliminate the effects of SA.

In practice, there are several additional sources of error other than SA that can affect the accuracy of a GPS-derived position. These include unintentional clock and ephemeris

---

[4] U.S. Department of Transportation and U.S. Department of Defense, *1992 Federal Radionavigation Plan*, DOT-VNTSC-RSPA-92-2/DOD 4650.5 (Springfield, Virginia: National Technical Information Service, January 1993).

[5] SPS accuracy is normally represented using a horizontal 2 drms measurement, or twice the root mean square radial distance error. Normally, 2 drms can be represented graphically as a circle about the true position containing approximately 95 percent of the position determinations. The definition of 2 drms and other positioning accuracy definitions are discussed in greater detail in Appendix D.

[6] SEP, or spherical error probable, represents an accuracy that is achievable 50 percent of the time in all three dimensions (latitude, longitude, and altitude). PPS accuracy is normally represented in this manner. The 2 drms PPS specified accuracy value is 21 meters SEP, as shown in Figure C-7 in Appendix C.

errors, errors due to atmospheric delays, multipath errors, errors due to receiver noise, and errors due to poor satellite geometry. Each of these error sources is discussed in Appendix C.

Even before the implementation of SA in 1990, many potential GPS users envisioned a need to improve the accuracy of the system, as well as some of its other specified characteristics. These other operational characteristics include integrity, availability, continuity of service, and resistance to radio frequency (RF) interference. These important concepts are defined and discussed in Appendix C.

Many techniques and technical systems designed to improve the capabilities of the basic GPS have been proposed, are under development, or are already in operational use. These techniques range from the use of GPS in a differential mode, to software and hardware improvements for GPS user equipment, to the integration of GPS user equipment with other navigation/positioning systems. Examples of each of these major areas of GPS enhancement are discussed in Appendix C.

2

GPS Applications and Requirements

INTRODUCTION

GPS specifications were originally developed by the DOD (Department of Defense) in the late 1960s with the primary objective of satisfying military navigation requirements. A secondary objective was to provide a separate, less accurate signal for both military and civilian use. This signal, described in Appendix C, and known as the Standard Positioning Service (SPS), was intentionally degraded in accuracy (100 meters, 2 drms) to avoid its exploitation by potentially unfriendly users.

As the GPS satellite constellation expanded and was eventually completed in 1993, the use of the freely available SPS signal for civil applications also continuously expanded. GPS is now used for positioning, navigation, and timing applications in a number of civil and commercial activities related to aviation; maritime commerce and recreation; land transportation; mapping, surveying, and geodesy; scientific research; timing and telecommunications; and spacecraft. Each of these broadly defined civilian user categories, along with military applications, is discussed in this chapter.

Many of the innovative civilian applications that this chapter will address were not foreseen by the original designers and developers of GPS and cannot be accomplished without augmenting and/or enhancing the stand-alone capabilities of the system as currently configured. As a result, differential correction methods and user equipment integrated with other positioning technologies, as described in Appendix C, have been utilized to meet the requirements of many of these applications. Within this context, there have been no requirements imposed on the basic GPS by civilian users to date beyond the assurance that the basic SPS signal-in-space will remain freely available at its currently defined accuracy level.[1] Users have taken this signal and adapted it to their applications. The basic GPS has therefore become a "dual-use" system, which is still designed to meet the requirements of only a single user, the Department of Defense.[2]

---

[1] This official U.S. government policy is currently reiterated every 2 years in the *Federal Radionavigation Plan*.

[2] The term "dual-use" usually refers to use by both the military and civilians.

Although the continued existence of GPS as a dual-use system clearly requires some trade-offs between civilian utility and national security, the NRC committee has concluded from its deliberations that because GPS provides tremendous benefits to both civilian and military users, as the remainder of this chapter will clearly illustrate, it should firmly remain a dual-use system. From the committee's perspective, recognition that GPS is truly a dual-use system brings with it the responsibility of meeting the requirements of all users to the highest degree possible. This implies that the system must be designed to the specifications of both civilian and military requirements. Many nonmilitary users of GPS have requirements that have been validated by standard-setting bodies and federal agencies that can now only be achieved through the additional cost of differential GPS (DGPS). Because human safety is an important consideration for many of these applications, a specified level of accuracy is not the only requirement. Integrity, availability, and resistance to RF (radio frequency) interference (both intentional and unintentional), as defined in Appendix C, are of significant importance as well. The sections that immediately follow discuss these requirements for each user category.

The task given to the NRC committee by Congress also recognized the dual-use nature of GPS and the trade-offs that exist between civil and military utility when it asked the following questions: "What augmentations and technical improvements to the GPS itself are feasible and could enhance military, civilian, and commercial use of the system?"; and, "What are the implications of security-related safeguards and countermeasures for the various classes of civilian GPS users?" These questions are examined in the remainder of the chapter by determining the challenges that currently exist for full utilization of GPS in each user community, including challenges that are related to Selective Availability (SA) and Anti-Spoofing (A-S). Although some of these challenges relate to the limitations of associated technologies and technology policies, findings in this chapter reveal that the biggest challenge for most users is meeting the requirements of a given application through augmentation of the GPS SPS. It stands to reason, therefore, that improving the basic capabilities of GPS and the freely available SPS signal will enhance the ability of civilian users to meet their requirements more easily, more cost-effectively, and in some cases, without augmentation or enhancement from DGPS or other positioning technologies. Improvements to the basic GPS can be made that will improve the military's ability to meet its requirements as well. Specific technical recommendations that would achieve this goal and address the tasks assigned to the NRC committee are discussed in detail in the next chapter.

GPS MILITARY APPLICATIONS

Although the overall use of GPS in the civilian sector has grown much faster than military usage, the system was designed with military requirements in mind, and the importance of the system to national security has not diminished. GPS is more accurate than any other radionavigation or positioning technology developed by the DOD, and is beginning to replace all other operational systems.[3] The coalition military forces demonstrated the effective use of GPS for many of these proposed applications during the Persian Gulf War, despite the fact that the GPS constellation consisted of only 16 satellites at the time. This limited three-dimensional coverage of the Persian Gulf region to 18 hours per day. Another limiting factor was the small number of P-code military receivers in the DOD inventory at the time of the conflict. This prompted a National Command Authority decision to turn SA to zero during the war and led to the DOD's purchase of thousands of civilian GPS receivers, which became known as "sluggers."[4] In addition to this official procurement, many units and individuals deployed to the Persian Gulf ordered their own GPS receivers directly from vendors and manufacturers.[5]

Current and Future Applications and Requirements

The most common use of GPS during the Persian Gulf War, and perhaps the most critical, was for land navigation. U.S. Army tanks and infantry relied heavily on GPS to avoid getting lost during movements to various destinations in the featureless desert. GPS also was used by coalition forces for en route navigation by aircraft, helicopter search and rescue, marine navigation, and even munitions guidance in the case of the U.S. Navy's Standoff Land Attack Missile (SLAM).[6]

The use of GPS for precision-guided munitions such as the SLAM will increase in the future. The U.S. military currently has, or is developing, eight additional types of precision land attack weapons that utilize GPS integrated with inertial navigation systems for mid-course guidance.[7] Another important GPS application under consideration is the

---

[3] The DOD plans to phase out use of Loran-C and Omega in 1994, Transit in 1996, and land-based navigation aids by 2000, depending on the progress of GPS installation and integration. Civilian use of these systems, however, may continue. Source: Radionavigation System Users Conference held in Washington D.C. on November 9-10, 1993, (unpublished).

[4] The "slugger" or Small Lightweight GPS Receiver is a Trimble Navigation TRIMPACK, three-channel receiver that utilizes the L_1, C/A-code to provide three-dimensional navigation capability. More than 10,000 receivers were purchased by the DOD from Trimble Navigation and other receiver manufacturers during the Persian Gulf War.

[5] Bruce D. Nordwall, "Imagination Only Limit to Military, Commercial Applications for GPS," *Aviation Week & Space Technology*, 14 October 1991, p. 60.

[6] Joseph Wysocki, "GPS and Selective Availability — The Military Perspective," *GPS World*, July/August 1991, pp. 38-43.

[7] These eight weapons include: the Tomahawk Block III and IV cruise missile; the Tri-Service Stand-Off Attack Missile; the Joint Direct Attack Munition; the Joint Stand-Off Weapon; the GBU-15 precision glide bomb; the AGM-130, a powered version of the GBU-15; and finally, the ATACMS ballistic missile. Source: J.G. Roos, "A Pair of Achilles Heels: How Vulnerable to Jamming are U.S. Precision-Strike Weapons?" *Armed Forces Journal International*, November 1994, p. 22.

use of GPS for the precision delivery of cargo by parachute or paraglider. For this application, GPS must be capable of providing steering commands to a reefing system to steer the parachute or paraglider to the desired landing point.

Combat search and rescue is another important function for which the use of GPS is increasing. Although GPS is already used for navigation by helicopters and other aircraft involved in combat search and rescue, it also will be used in the future to determine the exact location of downed aircrew members. By combining GPS with space-based communications capabilities, individuals can be found quickly, saving lives, time, and money. Communications capabilities would allow the location of aircraft, helicopters, and tanks to be monitored in real time, reducing casualties by friendly fire. Further, if GPS and communications capabilities are combined with guidance systems, unmanned aerial vehicles could be used for surveillance of target areas.

Tables 2-1 through 2-3 represent an extensive list of the military's positioning and navigation applications and their requirements.

Challenges to Full GPS Utilization

Accuracy and Integrity

The shaded cells in Tables 2-1 through 2-3 point out positioning and navigation requirements that cannot be met with the current 16 meter (SEP) specified accuracy[8], or 8-meter (CEP) derived accuracy[9], of the GPS PPS (Precise Positioning Service). Presumably, many of these requirements are currently being met by other guidance systems, such as highly accurate inertial navigation systems and terminal seekers, and other radionavigation systems, such as the microwave landing system. If these applications were to rely on GPS alone in the future, their accuracy requirements could only be met with some form of DGPS or a significantly improved PPS.

Some of the aviation applications listed in Table 2-1 also have specified integrity requirements. These requirements cannot be met with the PPS as currently configured.

---

[8] SEP, or spherical error probable, represents an accuracy that is achievable 50 percent of the time in all three dimensions (latitude, longitude, and altitude). PPS accuracy is normally represented in this manner. The 2 drms PPS specified accuracy value is 21 meters SEP, as shown in Figure C-7 in Appendix C.

[9] CEP, or circular error probable, represents an accuracy that is achievable 50 percent of the time in two dimensions (latitude and longitude). Most military accuracy requirements are defined in this manner. CEP, and other positioning accuracy definitions are discussed in greater detail in Appendix D.

Table 2-1   Military Aviation and Precision-Guided Munitions Applications and Requirements[a]

	Application	Accuracy	Integrity		Resistance to RF Interference
			1 minus P_{HE} times P_{MD}[b]	Time to Alarm	
Aviation[b]	Low-level Navigation and Air Drop	50.0 m (2 drms)	0.999	10 sec	High
	Non-precision Sea App/Landings	12.0 m (2 drms)	0.999	10 sec	High
	Precision App/Landings Unprepared Surface	12.5 m (2 drms)	0.999	6 sec	High
	Precision Sea App/Landings	0.6 m (2 drms)	0.999	6 sec	High
	Amphibious and Anti-submarine Warfare	50.0 m CEP	Not specified	Not specified	High
	Anti-air Warfare	18.1 m CEP	Not specified	Not specified	High
	Conventional Bombing	37.5 m CEP	Not specified	Not specified	High
	Nuclear Bombing	75.0 m CEP	Not specified	Not specified	High
	Close Air Support/Interdiction	9.0 m CEP	Not specified	Not specified	High
	Electronic Warfare	22.5 m CEP	Not specified	Not specified	High
	Command, Control & Communications	37.5 m CEP	Not specified	Not specified	High
	Air Refueling	370.0 m CEP	Not specified	Not specified	High
	Mine Warfare	16.0 m CEP	Not specified	Not specified	High
	Reconnaissance	18.1 m CEP	Not specified	Not specified	High
	Magnetic and Gravity Survey	20.0 m CEP	Not specified	Not specified	High
	Search & Rescue and Medical Evacuation	125.0 m CEP	Not specified	Not specified	High
	Mapping	50.0 m CEP	Not specified	Not specified	High
Precision-guided Munitions	Precision-guided Munitions	3.0 m CEP	Not specified	Not specified	High

a.   Availability and continuity of service requirements are not specified for military aviation and precision-guided munitions applications.

b.   This measure relates the probability that a hazardously misleading error will occur (P_{HE}) and the probability that this error will go undetected (P_{MD}).

c.   Peacetime requirements for the en route through Category I approach and landing phases of flight are identical to FAA requirements.

Table 2-2 Naval Applications and Requirements[a]

	Application	Accuracy	Resistance to RF Interference
En route Navigation	Pilotage & Coastal Waters	72.0 m CEP	High
	Inland Waters	25.0 m CEP	High
	Open Waters	2400.0 m CEP	High
	Rendezvous	380.0 m CEP	High
	Harbor	8.0 m CEP	High
Mine Warfare	Swept Channel Navigation & Defensive Mining	16.0 m CEP	High
	Offensive Mining	50.0 m CEP	High
	Anti-mine Countermeasures	< 5.0 m CEP	High
	Geodetic Reference Guide (WGS-84)	128.0 m CEP	High
Special Warfare	Airdrop	20.0 m CEP	High
	Small Craft	50.0 m CEP	High
	Combat Swimming	1.0 m CEP	High
	Land Warfare & Insertion/Extraction	1.0 m CEP	High
	Task Group Operations	72.0 m CEP	High
Amphibious Warfare	Beach Surveys	185.0 m CEP	High
	Landing Craft	50.0 m CEP	High
	Artillery & Reconnaissance	< 6.0 m CEP	High
Surveying	Hydrographic	< 5.0 m (2 drms)[b]	High
	Ocean & Geophysical Deep Ocean	90.0 m (2 drms)	High
	Oceanographic	100.0 m (2 drms)	High

a.   Availability, integrity, and continuity of service requirements are not specified for naval applications.
b.   This requirement can currently be met with data post-processing.

Table 2-3 Military Land Applications and Requirements[a]

	Application	Accuracy	Resistance to RF Interference
	Chemical Warfare	100.0 m CEP	High
Engineer	Mine Neutralization	100.0 m CEP	High
	Mine Dispensing & Gap Crossing	50.0 m CEP	High
Field Artillery	MLRS	20.0 m CEP	High
	Howitzer	17.5 m CEP	High
	Mortars	50.0 m CEP	High
	Fist-V & Forward Observer	30.0 m CEP	High
	Artillery and Mortar Radar	10.0 m CEP	High
	Infantry & Armor[b]	100.0 m CEP	High
	Missile Munitions	93.0 m CEP	High
	Signal	15.0 m CEP	High
	Special Operations Forces	30.0 m CEP	High
	Intelligence Electronic Warfare	20.0 m CEP	High
	Ordnance	84.0 m CEP	High
Air Defense Artillery	Patriot	10.0 m CEP	High
	Hawk	40.0 m CEP	High

a.   Availability, integrity, and continuity of service requirements are not specified for military land transportation applications.
b.   The Infantry & Armor category also includes transportation, soldier support, military police, and quartermaster.

Anti-Jam and Anti-Spoof Capability

Although the "Resistance to RF Interference" column in Tables 2-1 through 2-3 does not include quantitative values, a high level of resistance to RF interference is a critical requirement for most military applications.[10] For the military, the primary interference concerns are deliberate jamming and spoofing by an adversary or by our own forces. In future conflicts, a potential enemy also will be utilizing the capabilities of GPS and DGPS against U.S. and allied military forces. In order to deny this use, friendly forces must have the ability to eliminate an adversary's use of GPS signals without impacting the effectiveness of their own user equipment. This dictates that military GPS receivers also must be capable of continued operation in an environment populated with both U.S. and enemy jammers. Therefore, GPS-based navigation systems used on aircraft, ships, land vehicles, and precision-guided munitions must possess one or more of the following capabilities:

---

[10] Quantifiable values for resistance to RF interference are given in decibels (dB), and relate to the ratio of jammer power to signal power (J/S). These values are very specific to a given mission and operational environment, making a generic J/S requirement for a given application difficult to determine.

- have sufficient jamming-to-signal ratio strength to navigate through the jamming environment successfully;

- be able to null out the jamming signal; and/or

- have an alternative to GPS for navigating through the jamming environment.

The military currently relies on its SA and A-S (Anti-spoofing) security procedures to deny full GPS accuracy to the enemy while maintaining the use of a highly accurate spoof resistant signal. Anti-jam antennas and antenna electronics also are deployed on many weapons systems to provide increased jam resistance, and integrated GPS/inertial navigation systems provide a means of navigating to a target in spite of successful jamming. None of these procedures and technical measures, however, can be considered the final solution to the military's requirement to simultaneously use GPS and deny its use to the enemy. A number of candidate improvements in this regard are presented in the next chapter.

Findings

The GPS PPS meets most of the military's positioning and navigation accuracy requirements, although some applications require accuracy and integrity that is beyond the capability of the PPS as currently configured.

The anti-jamming and anti-spoofing capabilities of military GPS user equipment are critical to successful mission completion in a battlefield environment characterized by both U.S. and enemy spoofers and jammers.

GPS AVIATION APPLICATIONS

Despite the fact that investigations into the use of satellites for civil aviation applications have been conducted for over 20 years, the concept was not considered financially or technically feasible until the development of GPS.[11] Instead, a large number of ground-based radionavigation systems have been relied upon around the world for air navigation services, and ground-based air traffic controllers have utilized radar, voice position reporting, and visual sightings for aircraft surveillance.[12] It now appears that a

---

[11] Federal Aviation Administration (FAA), *FAA Satellite Navigation Program Master Plan*. FAA Research and Development Service, Satellite Program Office (ARD-70), 15 February 1993, p. 2.

[12] Existing ground-based radionavigation systems include NDBs (Non-Directional Beacon), VORs (VHF Omni-directional Range), VOR/DMEs (VORs with Distance Measuring Equipment), TACANs (Tactical Air Navigation), and VORTACs (combined VORs and TACANs). Other systems include the Instrument Landing System (ILS), used for precision approach and landing, and Loran-C and Omega, both of which are used for en route navigation. Each of these systems is described in detail in the 1992 *Federal Radionavigation Plan*.

Global Navigation Satellite System (GNSS), based on GPS and additional satellite augmentations, could eventually replace most of these ground-based systems.

Current and Future Applications and Requirements

Civilian pilots have been utilizing GPS in uncontrolled airspace for applications such as crop dusting, aerial photography and surveying, search and rescue, and basic point-to-point navigation for some time.[13] On June 9, 1993, the Federal Aviation Administration (FAA) approved GPS for supplemental use in the domestic, oceanic, terminal, and non-precision approach phases of flight in controlled airspace as well. This supplemental use required that another navigation source, such as a ground-based radio aid, must still be monitored while using GPS as the primary system. Once initial operating capability was declared for GPS by the DOD and DOT (Department of Transportation) on December 8, 1993, the monitoring of another navigation system for integrity purposes became unnecessary, provided that the GPS receiver utilized meets the FAA's TSO C-129 criteria for Receiver Autonomous Integrity Monitoring (RAIM).[14] In addition, traditional navigation sources such as VORs and TACANs must still be operational, and their associated receiver equipment must be on board the aircraft as a backup. Several GPS receivers have already been certified under the FAA's TSO C-129 criteria.

The use of GPS as the primary means of navigation for the domestic en route through non-precision approach phases of flight will require better availability and continuity of service (reliability) than is currently available from the stand-alone system. Phase I of the FAA's Wide Area Augmentation System (WAAS), which is scheduled to be in place by 1997, will make this possible. Table 2-4 contains the quantitative performance requirements that the WAAS is being designed to meet.[15]

In the near future, the FAA hopes that GPS also will be used for Category I precision approaches. Precision approaches are required when the weather conditions at a given airport reduce the ceiling, or height of the base of a cloud layer, and the visibility, or the distance a pilot can see visually, to levels that are below non-precision approach criteria.[16] Phase II of the FAA's WAAS implementation, scheduled for completion in 2001, will improve GPS-derived accuracy enough to allow the system to be used for these types of approaches. This increased accuracy requirement, which was also derived from the WAAS request for proposal (RFP), is included in Table 2-4.

---

[13] In uncontrolled airspace, pilots are not in direct communications with air traffic controllers, are responsible for their own navigation, and must be able to avoid terrain and collisions with other aircraft visually.

[14] RAIM is discussed in the next chapter, and is further explained in Appendix C.

[15] Federal Aviation Administration. Wide Area Augmentation System (WAAS), Request For Proposal, DTFA01-94-R-21474.

[16] Category I approaches can be flown when the visibility is no less than 0.81 kilometers (0.5 miles), and the ceiling is no lower than 61 meters (200 feet).

Testing by the FAA and several contractors is currently underway to determine the feasibility of also using GPS to conduct Category II and III approaches and landings, and the results to date have been very promising. These approaches are flown when the weather conditions at an airport are even worse than those described previously for Category I.[17] As can be expected, the accuracy, integrity, and continuity of service requirements are stricter than those for Category I landing systems, and therefore, the concepts currently under development utilize local-area differential GPS augmentations, rather than the WAAS. The requirements for Category II and III, which were derived from the *Federal Radionavigation Plan* and existing International Civil Aviation Organization (ICAO) requirements for instrument landing systems (ILS), are listed in Table 2-4.[18]

GPS also shows promise for use in Traffic Alert/Collision Avoidance Systems (TCAS) and Automatic Dependent Surveillance (ADS) systems. TCAS is already used by U.S. airlines and by many airlines in Europe.[19] Testing of an updated TCAS, which broadcasts an aircraft's position and velocity derived from GPS on the existing Mode-S datalink, has proven to be more accurate than the existing system.[20] The requirements for this application are listed in Table 2-4.

ADS systems, which are still under study and development, would automatically broadcast an aircraft's GPS-derived position to the air traffic management (ATM) system via geostationary communications satellites in oceanic airspaces, and via terrestrial-based communications links in domestic airspace.[21] This would allow for more efficient ocean crossings than are currently possible using the existing ATM reporting system. ADS would also be useful in the domestic en route and terminal phases of flight, where current aircraft separation is primarily the responsibility of air traffic controllers who utilize secondary surveillance radars. ADS systems are also being considered for monitoring the land-based operations of an airport, such as aircraft taxiing, and service-vehicle collision avoidance. The requirements listed for ADS in Table 2-4, which are based on current radar-based surveillance requirements, should be considered preliminary because the FAA is in the early phases of studying how to use GPS in performing the surveillance function.

---

[17] For example, a properly equipped aircraft can fly a Category IIIB approach when the ceiling is below 15 meters (50 feet) and the visibility is between 50 and 200 meters. Source: Federal Aviation Administration, *FAA Advisory Circular No. 120-28C: Criteria for Approval of Category III Landing Weather Minima*, 9 March 1984.

[18] These requirements are currently under review and may be revised due to an emerging concept known as required navigation performance (RNP). See: R. J. Kelley and J. M. Davis, "Required Navigation Performance (RNP) for Precision Approach and Landing with GNSS Application," *Navigation: Journal of the Institute of Navigation* 41, no. 1 (1994): pp. 1-30.

[19] The current TCAS configuration uses a data link known as Mode-S to measure the vertical separation between two aircraft in close proximity to one another. Measurements that are determined to be too close by the TCAS software set off an alarm that warns the flight crew and allows them to take action.

[20] "FAA Redirects TCAS-3 Effort," *Aviation Week and Space Technology*, 27 September 1993, p. 37.

[21] The exact method of transmission in U.S. domestic airspace has not yet been determined.

Table 2-4 GPS Performance Requirements for Aviation Applications[a]

	Application	Accuracy (2 drms)	Integrity		Availability	Continuity of Service	Resistance to RF Inter-ference
			1 minus P_{HE} times P_{MD}[b]	Time to Alarm			
Navi-gation	En route Oceanic[c]	23.0 km	Not Specified	30.0 s	99.977%	Not Available	High
	En route to Non-Prec. App/Landing	100.0 m	$1 - 1 \times 10^{-7}$ per hour[d]	8.0 s	99.999%	$1 - 1 \times 10^{-8}$ per hour	High
	CAT I App/Landing	7.6 m	$1 - 4 \times 10^{-8}$ per app.	5.2 s	99.9%	$1 - 5.5 \times 10^{-5}$ per app.	Very High
	CAT II App/Landing	1.7 m (vertical)	$1 - 0.5 \times 10^{-9}$ per app.	2.0 s	Not specified	$1 - 2 \times 10^{-6}$ per 15 sec.	Very High
	CAT III App/Landing	0.6-1.2 m (vertical)	$1 - 0.5 \times 10^{-9}$ per app.	2.0 s	Not specified	$1 - 2 \times 10^{-6}$ per 15 sec.	Very High
Survei-llance	TCAS	14.4 m[e]	Not Specified	Not Spec.	Several days[f]	Essential equip.[g]	Installed equipment[h]
	Oceanic ADS	Not specified	Not specified	Not spec.	Not specified	Not specified	Not specified
	Domestic ADS	200.0 m[i]	Not specified	Not Spec.	99.999%[i]	Not specified	Very High
	Surface Surveillance	12.0 m (resol.)[j]	Not specified	Not spec.	99.87%[j]	Not specified	Very High

a.   Unless otherwise annotated, GPS aviation requirements were provided by the MITRE Corporation.

b.   This measure relates the probability that a hazardously misleading error will occur (P_{HE}) and the probability that this error will go undetected (P_{MD}).

c.   Source of en route oceanic requirements: U.S. Department of Commerce, National Telecommunications and Information Administration, *A Technical Report to the Secretary of Transportation on a National Approach to Augmented GPS Services*, NTIA Special Publication 94-30, November 1994, p. 12. It is likely that the accuracy requirement will become significantly more stringent in the future to allow tighter spacing between aircraft.

d.   This number is equivalent to 0.9999999 or 99.99999 percent.

e.   Based on current TCAS specifications.

f.   According to airline minimum equipment list (MEL) practice approved by FAA certification.

g.   Based on reliability certification for essential equipment.

h.   Must meet installed equipment test. Otherwise unspecified.

i.   Based on current radar surveillance.

j.   Based on Airport Surface Detection Equipment-3 specifications, which require the resolution of two targets separated by 12 meters.

Challenges to Full Utilization of GPS

Selective Availability and Anti-Spoofing

When SA dithering of the GPS signals is employed, the DGPS corrections required to circumvent the resulting accuracy degradation must keep up with the dithering rate. This is not a problem for local-area DGPS, since the local correction broadcast usually has a sufficient data rate to provide timely corrections. The space-based WAAS, however, broadcasts its differential corrections as part of the navigation message data carried by a GPS-like L_1 signal. SA has a negative effect on this signal format; the high correction data rate necessary to keep up with the SA dither rate constrains the flexibility of providing additional information on this navigation message.

SA also decreases navigation availability and integrity monitoring availability for SPS users because the ranging errors it introduces require better satellite geometry for the specified 100-meter level of navigation accuracy. This sometimes rules out the operational use of GPS, especially when there are failed satellites present, and significantly reduces the effectiveness of RAIM.[22]

The employment of A-S, which overlays the Y-code on L_2 rather than the P-code, denies the second frequency needed for real-time ionospheric correction to all but authorized PPS users. Without dual-frequency receivers on board aircraft, the WAAS needs to employ a large network of ground sites to collect ionospheric data, that will be interpolated by the user to estimate the ionospheric delay in the pseudorange measurements. The disadvantages of this constraint are a decrease in the vertical positioning accuracy of wide-area DGPS, and an increase in the size, complexity, and cost of the WAAS ground network.

Resistance to Radio Frequency Interference

A-S also limits an SPS receiver's ability to deal with RF interference from known sources such as the third harmonic of some UHF (ultra-high frequency) television channels and airborne VHF (very-high frequency) transmitters. Solutions to the potential problem of RF interference must be found if GPS is to become the primary navigation and surveillance system for aviation, and organizations such as the RTCA are actively studying the issue. Resistance to interference can be greatly improved through the use of dual-frequency receivers that can track the code on both L_1 and L_2 because it is unlikely that interference from a single source will simultaneously affect both frequencies. As discussed in Appendix G, access to the wider bandwidth of the P-code, which is approximately 20 MHz (versus 2 MHz for the C/A-code), also would increase resistance to interference and reduce vulnerability to multipath.

---

[22] An analysis of the effects of SA on RAIM was conducted for this study by the MITRE Corporation. The results are presented in the next chapter, and the full analysis can be found in Appendix F.

Operational Procedures

Currently GPS is being approved by the FAA to operate under the same procedures used for existing navigation aids. If GPS is to provide more efficiency to present operations, however, there have to be accompanying changes in technical infrastructure and institutional culture. The major benefits of GPS navigation and surveillance will only be achieved when its coverage and accuracy are exploited to enable aircraft to fly user-preferred flight paths with minimal command and control from air traffic controllers. The benefits and enabling factors for these new operational procedures are discussed below.[23]

Most instrument flights are constrained to specified paths that facilitate the air traffic management (ATM) system's human-controlled separation of aircraft. Since GPS-equipped aircraft will be able to fly any desired flight path with high accuracy, users (especially air carriers) can potentially gain significant fuel and time efficiencies by having the ability to fly the most advantageous routing from one destination to another while independently amending their flight path as necessary to avoid congestion and potential conflicts with other aircraft. In order to make this change in procedure possible, as a minimum, the following enabling factors will have to be in place:

- automation that can cope with numerous aircraft flight path crossings, unlike the present essentially linear flow of traffic;

- a changed ATM culture that accepts a high level of automation for conflict prediction and resolution, and allows more autonomy in the cockpit for route selection and aircraft separation;

- highly reliable flight management systems aboard all aircraft to ensure that the same airport and route information is available to each aircraft flying in the national airspace system;

- two-way data links that provide an interface between ATM and aircraft flight management systems for such purposes as automatic negotiation of flight clearances (with pilot approval) and updates to airport and air route databases; and,

- cockpit display of traffic information to allow all aircraft to provide self-separation and enhanced collision avoidance.

---

[23] More information on this concept, known as "free flight", can be found in the following document: RTCA, Inc., *Report of the RTCA Board of Directors' Select Committee on Free Flight* (Washington, D.C., 18 January 1995).

Findings

The implementation of the FAA's WAAS should enable all navigation requirements through Category I precision approach to be met with wide-area DGPS. Category II/III approaches and landings will still require local DGPS augmentations.

The presence of SA and A-S increases the cost and complexity of WAAS and limits the effectiveness of RAIM.

The full navigation and surveillance capabilities of GPS will not be realized until air traffic management procedures and related technical systems are revised and modernized. In addition, GPS requirements based on the simultaneous use of the system for both navigation and surveillance must be established.

Radio frequency interference with GPS signals could prove to be a significant problem for aviation applications. Techniques to mitigate its effects, such as the use of a second GPS frequency, must be explored.

MARITIME USE OF GPS

In general, mariners use GPS for either navigation or positioning, although GPS has recently been applied to surveillance applications as well. It is important to define these broad categories of use before discussing more specific marine GPS applications and their requirements.

Marine navigation can be defined as the process of planning, recording, and controlling the movement of a craft or vessel from one place to another. During this process, there are generally concerns regarding commerce, expediency in transport, human safety, and environmental protection. When a vessel is navigating, it is often in situations where it is committed to a course of action based on these concerns. This has led to specific requirements for accuracy, integrity, availability, and area of coverage.

Marine positioning usually refers to activities such as hydrographic surveying, locating underwater objects, or other activities on the water where a vessel is not traversing a path to a destination. As with marine navigation, marine positioning generally has well-defined accuracy requirements, but because of the amount of time on station, integrity requirements can often be relaxed. Because of the cost of the resources used in conducting some positioning operations, however, lack of availability can have a severe economic impact.

In an effort to avoid the economic and environmental costs of vessel collisions and groundings, many of the nations ports and harbors are being equipped with surveillance systems known as vessel traffic services (VTS), which monitor the course and speed of ships, just as the air traffic control system tracks the flight paths of aircraft. Some of these systems operate with personnel similar to air traffic controllers who monitor and advise ships and pilots. Others are more automated and rely on the ability of ships to monitor themselves.[24] In both cases, GPS and DGPS are used to provide accurate positioning information that is integrated with other positioning, communications, and computing technologies.

Current and Future Applications and Requirements

The navigational use of GPS has evolved slowly in the maritime community, due in part to the lack of continuous service available from GPS until initial operational capability was declared in December 1993. Commercial shippers are now beginning to equip their vessels with GPS for navigation, however, and in April 1994 the Coast Guard declared that a GPS receiver meets the requirements for carriage of electronic position fixing devices as prescribed under US CFR Title 33, Part 164, section 164.41. The U.S. Coast Guard also has established a DGPS network that will eventually provide coverage to all U.S. coastal areas, ports, harbors, and inland waterways. Some commercial shippers have begun to experiment with DGPS capability as well.

Use of GPS among recreational boaters is becoming widespread. The low cost ($400–$2,000) of marine GPS equipment has made it an attractive alternative to other systems such as Loran-C and Transit. Recreational use of DGPS, however, has been limited to applications, such as yacht racing, that require improved position and velocity information. The additional cost of the beacon receiver used to receive the Coast Guard DGPS corrections has limited the recreational use of DGPS.

Maritime navigational requirements are well documented in the *Federal Radionavigation Plan*.[25] It breaks the marine navigation problem down into several distinct phases that relate to different geographical considerations. These are oceanic, coastal, harbor/harbor approach, and inland waterway. The oceanic and coastal requirements have been derived from the limitations of systems that have been used for these phases of navigation for some time, such as celestial plotting techniques, Loran-C, and Transit, whereas the harbor/harbor approach requirements were developed through research on ship maneuvering and the human factors involved in piloting large commercial vessels. Table 2-5 lists the current GPS requirements for the oceanic, coastal, and harbor/harbor approach phases of navigation.

Because official inland waterway requirements have not yet been adopted, the values shown in Table 2-5 for this phase of navigation should be considered as tentative estimates. Recent field trials of the Coast Guard's DGPS service, however, have demonstrated sufficient accuracy to satisfy the Army Corps of Engineers inland waterway construction requirement of 6 meters (2 drms). It is likely that this same system also could satisfy inland waterway navigation requirements. The Coast Guard's goal for their DGPS service is to

---

[24] An example of the former type system would be the U.S. Coast Guard's ADS (automatic dependent surveillance) system now in use in Prince William Sound, Valdez, Alaska. The private-sector VTS being developed for Tampa Bay by Tampa Bay VIPS, INC., is a good example of the latter type.

[25] Section 2.4 Civil Marine Radionavigation Requirements, pages 2-24 through 2-34.

achieve 3-meter (2 drms) accuracy for these operations, and provide the needed integrity and availability for navigation as well.

In contrast to navigation, GPS became practical for many positioning applications as soon as there were a few hours of satellite coverage each day. The Coast Guard, for example, began positioning navigation buoys with DGPS in 1990, when there were only 12 hours of GPS coverage per day. Other applications include the positioning of offshore oil platforms by petroleum companies and hydrographic surveying conducted by the National Oceanic and Atmospheric Administration (NOAA) to develop nautical charts. These users often augment the GPS standard positioning service with DGPS services provided by the Coast Guard or private sector companies.

Requirements for positioning applications are not well documented. Generally, positioning applications strive to achieve the best accuracy possible within a user's practical limitations, which are often related to time and cost. A system that satisfies marine navigation requirements for accuracy often satisfies some marine positioning requirements as well. Many high frequency, very-high frequency, ultra-high frequency and microwave systems have been developed and successfully used over the years to provide high accuracy positioning information in specific geographic areas. With DGPS coming on line and meeting the harbor/harbor approach requirement of 8 meters (2 drms), however, the need for these other systems has waned.

Marine surveillance systems, such as Coast Guard and commercial VTS, require accurate velocity data in addition to accurate positioning information. The continuous broadcast of velocity from each ship in a given VTS coverage area will allow pilots and VTS operators to take evasive action when two or more ships are approaching the same location at a fast closure rate. DGPS currently yields velocity accuracy on the order of 0.1 nautical miles per hour, which is sufficient for this application.

Table 2-5 Requirements for Maritime Applications[a]

	Application	Accuracy (2 drms)	Integrity Time to Alarm[b]	Availability	Coverage	Resistance to RF Interference
Navigation	Oceanic	1800-3700 m (1-2 naut. mi.)	Not specified	99.0%	Global	Moderate
	Coastal	460 m (0.25 naut. mi.)	Not specified	99.7%	U.S. Coasts	Moderate
	Harbor/ Harbor Approach	8.0-20.0 m	6-10 s	99.7%	Harbors and Approaches	High
	Inland Waterway[c]	3.0 m	6-10 s	Not yet defined	Inland Waterways	High
	Recreational Boating[c]	10.0 m	Not specified	99.9%	Coasts and Inland Waterways Nationwide	Moderate
Surveillance	Vessel Traffic Services[d]	10.0 m	Not specified	99.9%	Local	Very High
Positioning	Resource Exploration	1.0-3.0 m	Not applicable	99.0%	Global	Moderate

a. Integrity (1 minus P_{HE} times P_{MD}) and continuity of service requirements are not defined for maritime applications. Other maritime GPS requirements originate from the *Federal Radionavigation Plan*, pp. 2-26 through 2-28 unless annotated otherwise.

b. Source of time-to-alarm requirements: U.S. Department of Commerce, National Telecommunications and Information Administration, *A Technical Report to the Secretary of Transportation on a National Approach to Augmented GPS Services*, p. 11.

c. These values are not firmly established requirements. They are estimated useful values determined by the committee.

d. Source of Vessel Traffic Services Requirements: D. H. Alsip, J. M. Butler, and J. T. Radice, *Implementation of the U.S. Coast Guard's Differential GPS Navigation Service* (Washington, D.C.: USCG Headquarters, Office of Navigation Safety and Waterway Services, Radionavigation Division, 28 June 1993).

Challenges to Full Utilization of GPS

Associated Technologies

The positioning and navigation capabilities of GPS and DGPS do not solve the user's problems by themselves. For coastal and oceanic navigation, a GPS position (latitude and longitude) can be directly plotted on a paper nautical chart in the traditional fashion. This procedure often limits the accuracy of a position solution, not because of GPS errors but because of chart errors and plotting errors. More precise charts and plotting methods are therefore required in order to take advantage of the accuracy of GPS and DGPS.

The current state of the art in marine navigation is the Electronic Chart Display Information System (ECDIS), which is capable of displaying information from nautical publications, electronic navigational charts, and navigation sensors simultaneously. The International Maritime Organization (IMO) has recently drafted an assembly resolution on performance standards for ECDIS. Final approval of this draft document at the next meeting of the IMO assembly in 1995 would represent the first step in replacing paper nautical charts with computer-generated electronic charts for commercial vessel navigation. Under current IMO regulations, all merchant vessels are required to carry and use up-to-date paper charts. Once ECDIS's built to IMO standards are in use, vessels will not be burdened by a requirement to maintain paper charts and will have superior navigation capabilities in coastal and harbor areas.

A key issue for the timely implementation of ECDIS is the availability of digital data for the production of Electronic Navigational Charts. NOAA has begun the process of developing digital databases for electronic nautical charts, but faces serious resource limitations in this endeavor.[26] Existing hydrographic surveys are often very old, and must be updated before accurate digital data can be developed from them. At today's rate of progress, NOAA expects that it will take 5 to 10 years to digitize paper charts of U.S. waters.[27] Until this task is completed, charts will continue to be a source of marine navigation error that cannot be overcome by the widespread use of GPS.

Selective Availability

Despite the fact that users who desire accuracy better than 100 meters (2 drms) can now get it from DGPS services such as the U.S. Coast Guard's, SA still has a negative impact on the marine use of GPS. For recreational boaters, who prefer not to spend additional money on DGPS-capable receivers, this is especially true. Loran-C, which is still the most popular marine navigation system, is frequently used by fishermen to return to previously known fishing grounds with an accuracy of 20 to 30 meters. If GPS cannot meet or better this capability, recreational boaters, who could represent a large market for GPS, will be reluctant to embrace it in their operations.

SA also has a negative impact on the ability of commercial ocean-going vessels to use GPS as the navigation sensor for automatic piloting equipment. Controlling a ship by

---

[26] NOAA's electronic chartmaking efforts are the focus of an NRC report titled: *Charting a Course Into the Digital Era: Guidance for NOAA's Nautical Charting Mission*, Marine Board, National Research Council (Washington, D.C.: National Academy Press, 1994).

[27] NRC, *Minding the Helm*, Marine Board, National Research Council (Washington, D.C.: National Academy Press, 1994), pp. 227-228.

autopilot is preferred to manual control during long voyages because it saves fuel and reduces crew workload. This equipment requires stable velocity inputs, which are unavailable from the SPS with SA present. Methods exist to smooth or limit SA errors, such as the integration of inertial navigation systems with GPS, but vessel operators may be unwilling to bear this additional cost burden.

The Coast Guard's DGPS service itself is also affected by the presence of SA. In order to keep up with the high rate of clock dither present with SA, the system's radiobeacons must broadcast differential corrections at a high update rate. These corrections then require most of the bandwidth available on the 283 KHz to 325 KHz signal. A slower correction rate would allow the broadcast of other safety-related information that may be critical to mariners in the coastal and harbor regions.

Integrity

Under current operational procedures, the GPS master control station (MCS) does not monitor the integrity of the SPS. An improperly operating satellite can be detected by observing errors in the broadcast of the Y-code, but it is possible for errors to exist in the C/A-code regardless of the status of the Y-code. Because of this situation, the Coast Guard has stated that DGPS radiobeacons would still be required even in the absence of SA. Other integrity issues for maritime DGPS users result from the potential lack of accurate electronic nautical charts used in ECDIS's as was discussed above.

Availability and Radio Frequency Interference

RF interference to both GPS and DGPS radiobeacon's are significant issues for the commercial maritime user because interference has a direct impact on signal availability. In the marine radiobeacon band (283 KHz to 325 KHz), atmospheric interference from electrical storms will occasionally interfere with operations. Vessels operating with additional sources of navigation information can cope with lapses in availability, but users of only GPS and DGPS cannot.

Findings

GPS and DGPS are now in use in the maritime community for a number of navigation, positioning, and surveillance applications.

The full benefit of GPS and DGPS will not be realized by maritime users until systems such as ECDIS's eliminate errors produced by inaccurate charts and incorrect plotting. Up-to-date digital hydrographic data is required for the electronic charts utilized by these systems.

The presence of SA affects the acceptance of GPS by recreational boaters and some commercial users, and limits the ability of the Coast Guard's DGPS service to provide important safety–related information to its users.

LAND TRANSPORTATION APPLICATIONS

The civil land transportation sector of the nation's economy has generally been slow to adopt high technologies from other sectors such as aerospace or electronics. Recently, however, this trend is beginning to change due to an increased focus on initiatives such as the Intelligent Transportation System (ITS), which adapt defense-related technologies for uses in the civilian community.[28] More specifically, land transportation applications of GPS are growing rapidly, triggered by ever cheaper and more sophisticated equipment, an accelerated maturation of technology, widely available differential augmentations, and competition for economic and environmental responsiveness. All modes of land transportation, including trains, trucks, automobiles, all-terrain vehicles, bicycles, back-country skiers, hikers, and even pedestrians, have applications in which safety, position-location, and navigation are important, and have users who are therefore willing to use low-cost GPS, DGPS augmentations, or other comparable systems.

Current and Future Applications and Requirements

The trucking and railroad industries are currently the dominant land users of GPS for vehicle location and navigation, in part for reasons of competitive advantage in meeting the needs of just-in-time manufacturers and goods distributors. As on-time delivery becomes increasingly important to U.S. manufacturers and distributors, the trucking and rail industries and the international freight industry will require the ability to locate not only their vehicles or shipping containers, but also the components of their cargo when it consists of divisible elements, such as the packages handled by United Parcel Service or Federal Express. This must be accomplished with ever-greater accuracy and in near real-time. The tentative quantitative requirements for these GPS applications are listed in Table 2-6.

One of the largest near-term markets for GPS will probably be for automobile and light truck navigation and position-location. This market can evolve in a number of ways, since the automobile is used for a variety of purposes. On-board GPS and CD-ROM map systems are already being utilized by several rental car agencies, and at least one major U.S. automobile manufacturer already offers a GPS-based navigation system to its customers as

---

[28] ITS was formerly known as the Intelligent Vehicle/Highway System (IVHS). The name was changed to recognize the multi-modal nature of transportation.

an option.[29] It also is estimated that over half a million automobiles owned and operated in Japan already carry a GPS-based automobile navigation system.[30]

Although the final systems architecture and standards for the nation's ITS program have yet to be determined, the FHWA (Federal Highway Administration) anticipates that GPS will be an important component.[31] Potential ITS applications for GPS, in addition to vehicle navigation and position-location, include collision avoidance and control, vehicle command and control, automated bus stops, automated toll collection, accident data collection, a number of commercial vehicle regulatory activities, and ITS infrastructure management. Tentative requirements for these applications are included in Table 2-6.

GPS can also be used for the automatic guidance of farm vehicles engaged in precision farming. Also known as prescription farming, or site-specific crop management, precision farming gives the farmer the ability to apply precise amounts of fertilizer and pesticide to exact field locations based on the type of crop planted and the soil composition, potentially improving both the efficiency and cost-effectiveness of these operations. The positioning and navigation accuracy required for precision farming, as shown in Table 2-6, can only be met with local-area DGPS.

Much of the growth in the so called low-end, personal GPS receiver market can be attributed to transportation-related recreation activities involving both vehicles and pedestrians. Examples include "off-roading" with four-wheel drive vehicles, back-country skiing, mountain climbing, bicycling, hiking, and even golfing.[32] For those activities in which the potential for "getting lost" is high, and search and rescue services are often required as a result, GPS is much more than a useful gadget; it is a potentially life-saving device.

---

[29] This system, known as Guidestar, is offered as an option in General Motor's Oldsmobile 88 model. It uses GPS as an accuracy monitor for a dead-reckoning and map-matching navigation system.

[30] Source of information: Personal conversation with Michael Swiek, Executive Secretary of the U.S. GPS Industry Council.

[31] The leaders of the two teams that have been awarded Phase II ITS contracts for continuation of architecture design are Rockwell International and Loral Federal Systems. It is too early in the design process to determine exactly what role GPS will play in either team's final architectures. Source of Information: personal conversation with Mr. Lee Simmons, National Architecture Team Leader for ITS, FHWA, 22 February 1995.

[32] Several golf courses in the United States have experimented with DGPS systems mounted on golf carts to provide golfers with exact distances to the pin based on their location on the course.

Table 2-6 Land Transportation Requirements[a]

	Application	Accuracy (2 drms)	Integrity Time to Alarm	Availability	Coverage	Resistance to RF Inter- ference
Railroad	Train Control	1.0 m	5 s	99.7%	Nation	High
ITS and Vehicle Navigation/ Position- Location	Highway Navigation and Guidance	5.0- 20.0 m	1-15 s	99.7%	Nation	High
	Mayday/Incident Alert	5.0- 30.0 m	1-15 s	99.7%	Nation	High
	Fleet Management (AVL/AVI)	25.0- 1500 m	1-15 s	99.7%	Nation	High
	Emergency Response	75.0- 100.0 m	1-15 s	99.7%	Nnation	High
	Automated Bus/Rail- Stop Annunciation	5.0 -30.0 m	1-15 s	99.7%	Nation	High
	Vehicle Command and Control	30.0 -50.0 m	1-15 s	99.7%	Nation	Very High
	Collision Avoidance, Control	1.0 m	1-15 s	99.7%	Local	Very High
	Collision Avoidance, Hazardous Situation	5.0 m	1-15 s	99.7%	Local	Very High
	Accident Data Collection	30.0 m	1-15 s	99.7%	Nation	Moderate
	Infrastructure Management	10.0 m	1-15 s	99.7%	Nation	Moderate
Hazmat	Vehicle or Cargo Locations	5.0 m	1 s	99.7%	Nation	High
Precision Farming	Automatic Vehicle Guidance[b]	0.05 m	5 s	99.7%	Local	High
Search & Rescue	Location Determination[c]	10.0 m	minutes	99.0%	Nation	High
Recreation	Off-road Vehicles, Hikers, Back-country Skiers, etc.[c]	5.0 m	minutes	99.0%	Nation	Moderate

a.   Integrity (1 minus P_{HE} times P_{MD}) and continuity of service requirements are not defined for land transportation applications. Source of other requirements, unless otherwise annotated: U.S. Department of Commerce, National Telecommunications and Information Administration, *A Technical Report to the Secretary of Transportation on a National Approach to Augmented GPS Services*, p. 9.

b. Precision farming requirements were derived from information provided by the U.S. Department of Agriculture and the Deere & Company, Precision Farming Group.

c. The values listed for these applications are not firmly established requirements. They are estimated useful values determined by the committee.

Challenges to Full GPS Utilization

The major challenges for most uses of GPS in the land transportation sector include the need to meet the desires and requirements of an ever-increasing number of creative applications and the need to do so with technologically integrated equipment that is affordable, reliable, and reasonably durable. Each of these challenges deserves further explanation.

Accuracy Versus Other Requirements

For many civil land transportation purposes, the issue of accuracy of location will be dominated by right-of-way and construction surveying needs, not by vehicle, cargo, or personnel position-location or navigation issues. Where positioning and navigation accuracy is important, it is often related to requirements such as availability and integrity. For example, it is important in the cities for both people and freight movements, and dispatch control, to have good accuracy resolution not compromised by loss of signal lock due to tall buildings or interference from other radio sources. Similarly, the problem of resolution to a few meters, essential in mountainous terrain for numerous applications such as avalanche search and rescue, and forest fire control, is made more difficult by terrain and foliage, which can mask GPS signals.

For trucking and shipping, where vehicle and cargo location, and fleet dispatch and management are important, it seems clear that availability and coverage may be a greater challenge than greater accuracy. For example, a truck traveling from Boston, Massachussets to Seattle, Washington will not need to be located to within 25 meters of its actual position throughout the entire trip; however, it will need to be located. Satellite or terrestrial-based DGPS augmentation techniques must be adopted that provide better availability and coverage to the entire nation, not just the densely populated areas.

Users who do perceive a need for higher navigation and position-location accuracy than is available from the GPS SPS can generally meet this requirement by utilizing one of several commercially provided DGPS services. Systems that provide the differential correction via FM subcarrier seem especially suited to land transportation users, although their networks do not yet cover all of the nation. Additional enhancements to GPS receivers, such as the integration of small solid-state inertial gyroscopes and accelerometers, can also improve accuracy and other performance characteristics. Those users with accuracy requirements in the 20-meter range, however, which includes the rapidly growing automobile navigation market, would not need to augment or enhance GPS with DGPS or inertial gyroscopes if SA were eliminated.

For future ITS applications of GPS, such as automatic vehicle control and collision avoidance, integrated systems that use inertial navigation units and differential corrections will be required to meet the stringent accuracy, integrity, and availability requirements placed on any system that is fundamental to public safety. Current legal limitations, which restrict the ability for private sector interests to provide "navigation" services as opposed to "positioning" services, however, may negate the ability to use private sector DGPS providers to help meet these requirements. Although the removal of SA would not allow the standard positioning service to meet these requirements either, it would improve the performance of both wide-area differential systems such as the WAAS, and local-area systems.[33]

The Cost of Integrated Systems

Land transportation seems to offer unlimited opportunities for integrating GPS and DGPS with other complementary technologies related to communications, scanning, and digital imaging. Example technical systems include cellular phones, on-board fax and computer resources, driver performance and alertness equipment, and vehicle operations sensors. It seems reasonable to suggest that private sector creativity will be able to devise these integrated systems that will likely form the core of the nation's future transportation systems such as ITS. A few words of caution, however, should be considered.

In order for these systems to be widely accepted by potential users, their cost must be considered modest; they must be easy to use; and the equipment itself must be durable, reliable, and essentially maintenance free. "Gadgets" that fail to meet the above criteria or compromise the operational safety of a vehicle will never be accepted voluntarily by users in the surface–transportation community, especially the competitive commercial vehicle market. In addition, integrated positioning and communications systems utilizing GPS and other technologies will be most widely accepted if they help to fulfill bonafide public- and private-sector customer needs in a cost-effective manner. Systems introduced to the marketplace because of technology push are not likely to achieve widespread success.

Findings

There is a tremendous market for land navigation and positioning systems that integrate GPS with other technologies such as digital communications systems, driver performance and alertness equipment, vehicle operations sensors, and CD-ROM-based digital mapping and applications software. These systems, however, will only become widely accepted if costs continue to drop, high levels of reliability can be maintained, and reasonable durability can be assured.

---

[33] Potential improvements to DGPS techniques as a result of the elimination of SA are discussed in the next chapter.

Improved integrity, availability, and resistance to RF interference are as important to many land transportation GPS users as defeating the accuracy degradation caused by SA.

GPS-based automobile navigation systems, which require accuracies in the 5- to 20-meter range, would no longer require DGPS if SA were eliminated and further improvements were made to the basic GPS as suggested in Chapter 3. The elimination of SA would also improve the performance of those DGPS systems required for higher-accuracy applications, such as collision avoidance, that are important to the future Intelligent Transportation System.

MAPPING, GEODESY, AND SURVEYING APPLICATIONS

Currently, the fields of mapping, surveying, and geodesy are being transformed by a number of new and innovative technologies, including Geographic Information Systems (GIS), high-resolution remote sensing, and GPS. Of these, GPS has had the most important and immediate impact because of both cost savings and accuracy improvements over previous positioning technologies and techniques. The single most powerful feature related to GPS, which is not true of traditional mapping and surveying techniques, is that its use does not require a line of sight between adjacent surveyed points. This factor is paramount in understanding the impact that GPS has had on the surveying and mapping communities.

Current and Future Applications and Requirements

GPS has been used by the surveying and mapping community since the late 1970s when only a few hours of satellite coverage were available. It was immediately clear that centimeter-level accuracy was obtainable over very long baselines (hundreds of kilometers). In the early 1980s, users of GPS faced several problems: the cost of GPS receivers; poor satellite coverage, which resulted in long lengths of time at each survey location; and poor user-equipment interfaces. Today, instantaneous measurements with centimeter accuracy over tens of kilometers and with one part in 10^8 accuracy over nearly any distance greater than 10 kilometers can be made. The cost of "surveying-level" receivers in 1994 ranged from $10,000 to $25,000, and these costs are falling rapidly. Practitioners are developing numerous new applications in surveying, such as the use of GPS in a kinematic mode to determine the elevation of terrain prior to grading it for a storm water basin.[34]

Traditional land surveying is increasingly being accomplished using GPS because of a continuous reduction in receiver costs, combined with an increase in user friendliness. This

---

[34] "Kinematic" GPS surveying is accomplished using a reference receiver and one or more moving remote receivers. The carrier-phase measurements observed by the remote receivers and the static receiver are used in an interferometric mode to allow the positions of the remote receivers to be determined to the centimeter level in real time. More information on carrier-phase (interferometric) GPS techniques can be found in Appendix C.

trend towards the use of GPS has enhanced the volume of survey receiver sales because land surveyors outnumber geodesists (control surveyors) by at least one order of magnitude.[35] This usage has also increased the accuracy and accuracy requirements of surveying in general.

GPS is also increasingly being used as the core technology for integrated mapping systems. These systems are usually mobile (e.g., a van, train, airplane or any other vehicle) and contain a combination of sensors. These sensors include vision or imaging systems, laser ranging and profiling systems, ground penetrating radars, and other navigation sensors such as inertial navigation units. GPS provides positioning data when satellites are visible, and other sensors provide the spatial location data required for map making. The inertial systems, and sometimes the vision systems, are used to interpolate between GPS positions through periods when GPS satellites are lost from the vehicle's field of view. These mapping systems provide the surveying and mapping community with powerful new ways of acquiring accurate and current digital data.

In general, the availability of higher GPS accuracy has influenced various mapping and surveying requirements for three reasons: (1) people want the latest and the best; (2) past requirements were in some cases dictated by the cost of acquisition; and (3) if higher accuracy can be obtained, multiple purposes can be satisfied. As an example of requirements changing as a function of new capability, consider a problem of facilities management which deals with the inventory of transportation features such as the location and attributes (type, condition, and so forth) of a guardrail along a highway. Previously, the location was "required" by transportation departments to be accurate to ± 6 meters (20 feet), which is generally the best that is possible from scaling or plotting on a 1/24,000 USGS quadrangle. Using differential techniques a GPS position can easily be obtained in a real-time, dynamic environment to within ± 1.5 meters (5 feet). Users now realize that if accuracies of ± 0.3 meters (1 foot) can be obtained (and they can), the length of the guardrail, in addition to its location, can be obtained so that if the guardrail needs to be upgraded or replaced, an accurate estimate of the cost is available. This kind of analysis is growing rapidly as GPS becomes understood and applied to various problems. Clearly, concepts of this kind are widespread in GIS applications in natural resource planning, environmental problems, civil infrastructure enhancements, an so on. Analogous examples can be given for surveying and geodesy.

Accuracy requirements for surveying applications are generally satisfied at this time. The quest for better and better accuracy will continue, but any reasonable distance can currently be measured, with significant care, to one part in 10^8. In each of the categories in Table 2-7, the most stringent accuracy requirements are adopted because of the potential for multipurpose applications.

---

[35] Land surveying usually ignores the curvature of the Earth (except in leveling) and assumes that the Earth's surface is a plane. Control surveying does not make this assumption and is generally performed with an accuracy an order of magnitude better than land surveying.

Table 2-7 Current and Future GPS Requirements for GIS, Mapping, Surveying, and Geodesy[a]

	Application	Accuracy (2 drms)	Integrity	Availability	Coverage
			Time to Alarm[b]		
	Geographic Information Systems (GIS)	1.0-10.0 m	Minutes	98%	Worldwide
	Photogrammetry	0.02-0.05 m	Minutes	98%	Worldwide
	Remote Sensing	0.1-20.0 m	Not specified	98%	Worldwide
	Geodesy	0.01-0.05 m	Hours[c]	98%	Sites Worldwide
	Mapping	0.1-10 m	Hours[c]	98%	Sites Worldwide
Surveying	Hydrographic	0.05-10.0 m	Hours[c]	98%	Sites Worldwide
	Topographic	0.01-0.5 m	Hours[c]	98%	Sites Worldwide
	Boundary	0.01-0.05 m	Hours[c]	98%	Sites Worldwide

a. Integrity (1 minus P_{HE} times P_{MD}), continuity of service, and resistance to RF interference requirements are not defined for mapping, survey, and geodetic applications. Source of other requirements, unless otherwise annotated: The Ohio State University, Center for Mapping.

b. Source of time-to-alarm requirements: A Technical Report to the Secretary of Transportation on a National Approach to Augmented GPS Services, p. 13.

c. The integrity of the positioning data for each of these applications is validated in post-processing.

Challenges to Full GPS Utilization

It is important to understand that nearly all accuracy requirements presently can be met using DGPS. However, the cost of meeting these requirements would decrease if various enhancements to the basic GPS itself were implemented. In particular, eliminating SA and/or A-S would drive down the costs of new applications. GIS applications would benefit most from the elimination of SA because many GIS requirements could potentially be satisfied by the accuracies obtained from stand-alone GPS with SA set to zero and with other potential accuracy improvements. All real-time, dynamic surveying and mapping applications would benefit from improved signal acquisition. Faster integer ambiguity resolution, important to real-time kinematic survey and mapping applications, would be achievable with a second frequency unencrypted by A-S. As discussed in Appendix G, access to the wider bandwidth of the P-code, which is approximately 20 MHz (versus 2 MHz for the C/A-code), also would increase resistance to RF interference and reduce vulnerability to multipath. Improving the GPS orbit information (ephemeris) available to SPS users also would have a significant impact on the surveying and mapping community as longer and longer baselines could be measured in real-time with centimeter accuracy.

Additional challenges to the use of GPS in the mapping, surveying, and GIS community deal with receiver cost, service and maintenance, user friendliness, and interfaces with other software and hardware. For example, at the Center for Mapping at the Ohio State University, researchers have developed a real-time positioning capability accurate to 1.5 centimeters with as few as five satellites in view. The system is interfaced with software developed by the construction service industry that displays "cuts and fills" on a screen so that an operator of earth-moving equipment can grade earth in a prescribed fashion. One problem with the overall system is the necessity to use two GPS receivers that currently cost $25,000 each. The competitive technology costs about $40,000. If the price of the GPS receivers falls to $10,000 each, however, the GPS technology will dominate the market. This is especially true because GPS offers coordinates in three dimensions without line of sight requirements. The rival technology (laser plane) is one dimensional and requires line of sight. Hence, the challenge to using GPS for earth moving is focused on software integration and the costs of receivers. This same scenario is applicable to many other potential GPS applications envisioned at the moment.

Findings

Greater geodetic accuracy for mapping and surveying will be pursued in part because of the challenge of obtaining it. A few applications, such as determining the position of the blades of earth-moving equipment in real time will demand increased accuracy. Most applications, however, will be enhanced by cost savings from quicker acquisition of the same data. The elimination of SA and A-S, and the use of dual-frequency user equipment can improve data acquisition time.

For surveying, the weakest link in the utilization of GPS, aside from SA and A–S, is the precision of the GPS satellite orbits. While improving ephemerides will not significantly enhance positioning over short baselines, they will have a noticeable impact over baselines greater than 50 kilometers.

GPS EARTH SCIENCE APPLICATIONS

One's ability to measure the Earth, including its atmosphere and its ocean surfaces, has been greatly enhanced by GPS. New departures in scientific endeavor and commercial enterprise have begun, and initial results are very promising.

Current and Future Applications and Requirements

Meteorology

In meteorology, GPS can be used to measure atmospheric water vapor. Water vapor is the principal mechanism by which moisture and latent heat are transported in the atmosphere and is therefore closely linked to weather and climate. As discussed in Appendix C, GPS signals are delayed by the ionosphere and neutral atmosphere as they travel towards the surface of the Earth. This delay can be estimated by a receiver observing the two principal GPS transmission frequencies. When combined with surface pressure data, the estimated signal delay can provide a measurement of wet delay, which in turn, can be converted into precipitable water vapor. GPS sensing of precipitable water vapor with millimeter accuracy has been demonstrated successfully. The use of this technique for weather forecasting is being explored, and has been proposed for climate research.

Another innovative use of GPS for meteorology is the new field of Earth-atmospheric occulation measurements. This technique uses a GPS receiver on a satellite in low-Earth orbit to track a GPS satellite as it sets behind the Earth. As the GPS signal passes through the edge of the atmosphere it is refracted, causing delay and Doppler shift, which is measured with millimeter accuracy by the spaceborne receiver. The index of refraction of the atmosphere can then be determined as a function of height. This index can then be analyzed to produce atmospheric temperature profiles and a measure of water vapor content. The first demonstration of this promising GPS application, which is also important to global change research, is scheduled to take place in 1995.

Oceanography

One importance of GPS to the field of oceanography is its potential ability to determine precise orbital parameters for the Topex/Poseidon satellite, which in turn, provides accurate radar altimetry of the ocean's surface. In general terms, Topex/Poseidon data improve in several ways as more precise orbital information becomes available. The issue is to separate orbital error from tides, general circulation, and gravity-field error. General circulation needs to be determined at the 1-centimeter level, a reasonably easy task with the GPS precise positioning service (PPS), but difficult, or perhaps even impossible, with other methods of orbit determination. Orbital error would no longer be a significant factor for all Topex/Poseidon data if orbits could be determined with an accuracy of $\pm$ 1 millimeter. Using the GPS PPS, this is a distant, although not unobtainable goal.

In the wider context of oceanography, one can assert that every time there has been a real improvement in navigation whole new fields of study have opened. GPS with SA set to zero provides a real improvement in navigation. Ocean-surface height measured by ships at sea, and the positioning of a tomographic lagrangian drifter also can be accomplished with useful accuracy.[36] Other oceanographic positioning applications, such as the location of objects on the ocean floor, which is essential for drilling and sampling activities, require real-time accuracy of about 10 meters.

Geodynamics

In geodynamics, GPS is used to study relative motions on the surface of the Earth. The tectonic plates of the Earth's outer layers move relative to each other at rates within the range of 1 centimeter per year to 20 centimeters per year. Many earthquakes occur along the plate boundaries, a recent example being the earthquake in Northridge, California, which occurred on January 17, 1994. A few permanent GPS reference stations provided important data for the early determination of the Northridge earthquake mechanism, which had a displacement on the order of 1 meter.

Arrays of permanent GPS stations, coupled with a few interferometric strain meters, can be used to study crustal deformation in the time intervals between earthquakes. This information could be used to estimate the varying amounts of seismic risk in, for example, different parts of the Los Angeles area. The risk assessment could be used to determine appropriate local variations in building codes, freeway and subway construction, and other public projects.

In a very speculative vein, *if* GPS arrays and associated strain meters reveal premonitory or precursory signals for earthquakes, and *if* the signals are detected early enough to provide meaningful warnings to a region's population and public authorities, *then* it would become important to measure these signals in as near real time as possible, that is, with minimal post-processing. Only time will tell whether GPS arrays will become useful in this very speculative vein. If not, the improved study and understanding of the deformation of the Earth's crust and of the rupture process of earthquakes will still provide ample reason to establish and operate permanent geodetic GPS arrays.

Airborne Geophysics

Many of the measurement tools used historically by Earth scientists for regional studies are not sufficiently accurate to model physical processes and improve the understanding of natural hazards and the distribution of nonrenewable resources. Physical barriers, such as inaccessibility by land due to hazardous terrain, and limited resources which prevent the surveying of large areas by conventional means, pose other difficulties.

Collecting data remotely from satellites or aircraft can overcome some of the sampling problems. Satellite missions, however, require long lead times between concept and realization, making airborne platforms an attractive alternative for regional Earth studies.

---

[36] Tomography is the use of acoustic travel time to infer changes in acoustic wave speed due to changes in sea temperature and composition. A tomographic lagrangian drifter is a neutrally buoyant buoy equipped to record the arrival of acoustic pulses for use in tomography studies.

Aeromagnetic surveys have been used for half a century with great effect, but airborne gravity and topographic mapping depend on the ability to determine the aircraft's motion and position.[37]

Positioning to 100 meters horizontally and 3 meters vertically is required to provide useful measurements of gravity and terrain. GPS and DGPS are ideal for such positioning and a combination of GPS and an inertial navigation system to provide the acceleration of the aircraft, could enable studies of dynamic changes in topography and gravity, such as the expansion of a volcanic dome caused by the injection of magma. Using GPS and radar altimetry to obtain precise gravity anomaly maps, the regional prospecting for ore bodies, salt domes (petroleum reservoirs), or large anticlines (big domes that contain petroleum) can also be accomplished quickly and economically.

Accuracy Requirements

In general, the Earth science applications described above require much better positioning accuracy than was ever anticipated or intended from GPS, as Table 2-8 clearly illustrates. However, any static GPS reference station equipped with a dual-frequency geodetic receiver can currently be positioned "absolutely" at the centimeter level with respect to the international terrestrial reference frame with less than 24-hours of data. Relative positions between stations at regional scales can be determined at the few-millimeter level with very short observation times. This capability is due to major improvements in GPS software packages and the availability of very precise satellite ephemerides (10-centimeter accuracy) determined by the International GPS Service for Geodynamics (IGS).

The ephemeris information available from the IGS can also be used for post-processed dynamic positioning applications. Moving platforms up to several hundred kilometers away from a fixed DGPS base station can achieve 10-centimeter to 20-centimeter positioning accuracy using corrections based on IGS ephemeris data. Satellite clock information distributed by the IGS also is helpful for mitigating SA effects in post-processing of position, particularly in airborne and oceanographic applications.

---

[37] Airborne Geophysics is the subject of a recent NRC report titled: *Airborne Geophysics and Precise Positioning: Scientific Issues and Future Directions*, Board on Earth Sciences and Resources, National Research Council (Washington, D.C.: National Academy Press, 1995).

Table 2-8 GPS Earth Science Requirements[a]

	Application	Accuracy (2 drms)	Integrity (time to alarm)
Static	Meteorology	0.001 m	Hours
	Oceanography - General Ocean Circulation Determination	0.01 m	Hours
	Geodynamics	$0.001 \text{ m} + 10^9 \times$ baseline length	Hours
Dynamic	Oceanography — real-time Positioning and Navigation	10.0-30.0 m	Not specified
	Airborne Geophysics	3.0 m vertical	Minutes

a.   Integrity (1 minus P_{HE} times P_{MD}), availability, continuity of service, and resistance to RF interference requirements are not available for the GPS Earth Science applications covered by this table. Other requirements were derived from input received from the appropriate scientific community.

Challenges to Full GPS Utilization

Meteorology

A third GPS radio frequency would be very helpful in atmospheric studies. Also, the presence of A-S greatly increases costs and limits the performance of many techniques due to loss of low-elevation angle data and signal-to-noise ratio, even when using dual-frequency codeless receivers.

Oceanography

In general, spacecraft orbits determined from GPS data with A-S off are superior to those determined by other means, with A-S on this is not the case. A successor mission to Topex/Poseidon could be designed with receivers that would work well in the presence of A-S and, essentially, overcome this obstacle. However, it has been estimated that the additional cost of adding a space-qualified PPS receiver to a satellite would be about $500,000.[38] Much of this cost stems from the security measures that are required for the proper handling of classified equipment.

For other types of oceanographic research, SA is the central challenge to the usefulness of GPS. The 10-meter to 30-meter accuracies required to navigate research vessels, position buoys, and locate objects on the ocean floor cannot be achieved using GPS

---

[38] W. G. Melbourne et al., "GPS Flight Receiver Program for NASA Science Missions — A Unified Development Plan," (JPL D-10489). Jet Propulsion Laboratory, 10 February 1993.

alone, unless SA is eliminated and other improvements are made to increase the accuracy of the SPS.

Geodynamics

Despite improved post-processing software and the use of differential GPS, the effects of A-S and SA degrade the results by 50 percent or more, primarily through the loss of the signal-to-noise ratio using dual-frequency codeless receivers. The loss can be partially recovered by replacing existing receivers that are a few years old with newer equipment. Significant savings in time and costs would occur, however, if this was not necessary.

Airborne Geophysics

SA has little effect on airborne geophysical applications when differential GPS and post-processing are utilized. As with geodynamic applications, however, the presence of A-S greatly reduces the signal-to-noise ratio available to dual-frequency receivers. The dynamic, high-multipath environment that exists for GPS receivers on aircraft makes codeless receivers especially vulnerable to losing lock on the L_2 signal and requires a lengthy reacquisition time. In lieu of code-tracking capability on L_2 or an alternative L-band signal, improvements to the tracking loops in codeless receivers could improve this situation.[39]

Findings

Using post-processed GPS orbits provided by the IGS network of differential reference stations, the effects of SA can be eliminated for most Earth science applications, and with the use of dual-frequency "codeless" receivers, centimeter-level positioning accuracies can be achieved.

The availability of a second GPS frequency for civil use with unencrypted code would greatly enhance many Earth science applications that require high-precision accuracy. Dynamic, high-multipath applications, such as airborne geophysics, would benefit from faster acquisition and more robust tracking. Applications such as remote atmospheric sensing require submillimeter precision in the carrier-phase observables, which may be achievable using a second unencrypted signal.

---

[39] The effects of SA and A-S on the use of GPS in airborne geophysics are discussed in more detail in the NRC report *Airborne Geophysics and Precise Positioning: Scientific Issues and Future Directions*, Appendix A: Effects of Selective Availability and Anti-spoofing.

GPS TIMING AND TELECOMMUNICATIONS APPLICATIONS

Because the pseudoranging method used by GPS to establish three-dimensional position locations requires a highly accurate time standard, the system is ideally suited for applications that require precision timing and precise time transfer. GPS pseudorange measurements are based on the transit time of a signal from the GPS satellite to the user. Thus, if the locations of both the satellite and the observer are known, the difference in the user-clock offset from that of the satellite can be readily determined. Furthermore, if the satellite clock is referenced to a standard such as Universal Coordinated Time (UTC), as is the case with GPS, the observer can then determine user-clock offset from UTC.[40]

Current and Future Applications and Requirements

The time-transfer community was one of the first to realize benefits from GPS, since a full satellite constellation is not required for most time-transfer methods. In fact, the most accurate method of time transfer to date, known as GPS common-view, relies on the ability of two users on the globe to observe the same GPS satellite simultaneously, despite a large geographic separation. GPS common-view is currently used by the 55 international timing centers that are charged with the task of maintaining International Atomic Time (TAI) and UTC throughout the world.[41] A chain of common-view observations also is used to link the widely separated sites that are part of the National Aeronautics and Space Administration's (NASA) Deep Space Network.[42] Other time-transfer methods that utilize a single GPS satellite, as well as methods that require observations from multiple satellites, are used for a number of scientific research activities that require precise time synchronization of equipment located in different laboratories.[43]

---

[40] UTC is often referred to as Greenwich Mean Time because it refers to the time of day in Greenwich, England (U.K.).

[41] The official international timing center in the United States is the National Institute of Standards and Technology (NIST) Metrology Laboratory in Boulder, Colorado. This facility, along with 53 others, keep time relative to the master facility at the *Bureau International des Poids et Measures* (BIPM) in France.

[42] The Deep Space Network (DSN) consists of three tracking stations located near Barstow, California; Canberra, Australia; and Madrid, Spain. These stations receive telemetry data from deep space missions such as Galileo, and send commands that control spacecraft navigation and operation. The three tracking stations are monitored by the DSN's control center at the NASA Jet Propulsion Laboratory in Pasadena, California.

[43] Single satellite time-transfer methods in addition to common-view include GPS direct and clock flyover. Methods using multiple satellites include Enhanced GPS, GPS used as Very Long Baseline Interferometry (VLBI), and Geodetic Positioning Time Transfer. For more information on these methods see: David Allen, Jack Kusters, and Robin Giffard, "Civil GPS Timing Applications," in *Proceedings of ION GPS-94: 7th International Technical Meeting of the Satellite Division of the Institute of Navigation* (Salt Lake City, Utah, September 1994), pp. 25-32.

GPS is also increasingly utilized by many telecommunications companies to synchronize their land-based digital telecommunications networks.[44] Most often, these users compare a reference clock directly to GPS time by viewing one or more satellites, rather than transferring time from one reference clock to another. AT&T, in particular, now uses GPS to maintain time synchronization throughout its long distance telephone system,[45] and an international digital telecommunications system that uses a GPS-based timing system began operating in Moscow in 1991.[46] As synchronous fiber optic networks such as SONET's increase in size and complexity, GPS time synchronization may replace the more common practice of using land lines to disseminate timing information from a small number of land-based clocks.[47]

The "Stratum n" performance level hierarchy, developed by the American National Standards Institute (ANSI) T1 Committee on Network Synchronization Methods and Interfaces, specifies the requirement for synchronization. At the present, the one to four Stratum performance levels (with one being the most stringent) could be satisfied by the long-term frequency stability available from the GPS standard positioning service.[48] The ANSI T1 requirements are listed in Table 2-9.

Precise GPS timing also has the potential to significantly improve mobile cellular communications. Currently most cellular telephone networks are subject to transmission degradation as a call is transferred from one cell's channel to another, but if all of a network's cells used the same channel, this problem would be eliminated. This can be accomplished by providing each cell with a unique code rather than a unique frequency using a technique known as Code Division Multiple Access (CDMA).[49] Major CDMA manufacturers have recognized GPS as an effective way to provide the precise time synchronization required by their systems.[50] Timing accuracies similar to those required for digital networks are sufficient for this application.

---

[44] Information presented in this section on the use of GPS by the telecommunications industry, unless annotated otherwise, is based on the following report: Eric A. Bobinsky, *GPS and Global Telecommunications: A Summary Briefing Prepared for the National Research Council Committee on the Future of the Global Positioning System* (Washington, D.C., 29 July 1994).

[45] E. Krochmalny, "GPS Synchronizes the Lines," *GPS World*, May 1992, p. 39.

[46] M. J. Toolin, "GPS in a Russian Telecommunications Network," *GPS World*, June 1992, pp. 28-34.

[47] SONETs, or Synchronized Optical NETworks, were originally proposed by Bellcore, and are now becoming the worldwide standard format for optical transmissions. The term "synchronous" highlights the fact that a SONET is aligned in time with respect to a common timing source.

[48] There are currently no ANSI T1 "Stratum n" requirements for absolute timing accuracy. The absolute timing accuracy specification for the GPS SPS is 340 nanoseconds relative to UTC.

[49] Code Division Multiple Access is the same technique that allows a GPS receiver to distinguish one satellite from another despite the fact that they all use the same frequency.

[50] U. H. Werner, "Improving Mobile Communications with GPS," *GPS World*, May 1993, pp. 40-43.

Cellular signals are also subject to the local conditions in each cell that may vary from cell to cell, such as weather or landform geometry. By putting GPS positioning capability in the mobile receiver and by transmitting the position information to the mobile control and operations center of the mobile system, the network control operations could determine user location and travel direction. With this information available, the network controller can provide optimal hand over as well as real-time dynamic performance optimization for each location. A typical communications cell ranges from a few tens of meters to over a hundred square kilometers, so a positioning accuracy of a few hundred meters will suffice. When dealing with small, oddly shaped cells, however, or when trying to map signal and propagation characteristics within a complex area such as an "urban canyon," accuracy on the order of a few meters in three dimensions may be required. These general values for positioning accuracy have not yet been defined as requirements, and therefore are not included in Table 2-9.

In the future, many information services may require "time-of-day" information to a much higher degree of accuracy than is typical of today's services. Examples include universal personal communications services and broadband integrated services digital networks which may require a high degree of time-of-day precision in order to interface with several different types of communications systems to transmit tremendous amounts of digitally packeted information.[51] Timing accuracies of 100 to 300 nanoseconds relative to UTC will likely be required for these services.

Table 2-9 Timing and Telecommunications Requirements[a]

	Application	Accuracy[b]		Reliability[c]
		Time	Frequency	
Common-View Time Transfer	NASA Deep Space Network	1 ns	1×10^{-15}	Not specified
	BIPM for TAI and UTC	1 ns	1×10^{-14}	Not specified
	International Timing Centers	0.1-1 ns	1×10^{-14}	Not specified
	NIST Global Time Service	10 ns	1×10^{-14}	Not specified
Time Synchronization	Power Industry	10 ns	Not available	High
	ANSI T1 Stratum 1	Not specified	1×10^{-11}	High
	Time-of-Day Services	100-300 ms	Not specified	High

a.   Source of requirements for common-view time transfer and power industry time synchronization: David
      Allen, Jack Kusters, and Robin Giffard, "Civil GPS Timing Applications," p. 28. Source of time-of-day

---

[51] Iridium, Orbcomm, Globalstar and other proposed low-Earth orbit (LEO) satellite communications systems are all examples of UPC services. Broadband integrated services digital networks, are digital telephone lines capable of transmitting data, voice, graphics, and video information at a rate much faster than modems.

requirement: Eric A. Bobinsky, "GPS and Global Telecommunications." ANSI Stratum 1 requirements provided by Mr. Bruce M. Penrod of TrueTime, Santa Rosa, CA.

b.   The timing accuracies listed include both time relative to UTC in nanoseconds (ns) or milliseconds (ms), and long term frequency stability measured over one day, except for the ANSI Stratum 1 long term frequency stability, which is measured over any time interval greater than 1000 seconds.

c.   For commercial time synchronization applications, "reliability" corresponds to overall system reliability as explained in this section, not the continuity of service requirement applicable to GPS aviation applications.

Challenges to Full GPS Utilization

Time Transfer

For GPS time-transfer applications, the challenge of mitigating the effects of SA's clock dithering in order to improve accuracy appears to have been met. Methods to filter SA-induced noise have allowed time transfers to occur using C/A-code receivers, which achieve accuracies of better than 1 nanosecond relative to UTC, and long-term frequency stabilities of better than 1×10^{-14}.[52] Laboratories responsible for the world's primary time standards, such as NIST's metrology laboratory in Boulder, Colorado, are hoping to conduct time transfers with this type of accuracy on a routine basis. These accuracies are required in order to maintain standards that are two orders of magnitude better than the timing accuracies required by industry.

If errors from SA are removed and ionospheric errors are minimized by using dual-frequency receivers, clock and ephemeris errors become dominant. Improvements to the GPS space and ground control segments will be required in order to reduce these errors.[53]

Time Synchronization

For the telecommunications industry, requirements such as integrity and availability fall under the general category of overall system reliability. Communications and information service providers will not rely on any technical system that does not guarantee them the ability to satisfy the needs of their customers on a continuous 24 hour-a-day basis. Many potential users of GPS in the telecommunications industry feel that GPS, as currently configured, cannot provide this level of reliability. As with many other GPS applications, the absence of SPS integrity monitoring is unacceptable to many in the telecommunications industry. These potential users have expressed a desire to have GPS performance monitoring data available to them in real time in order to feel comfortable with its reliability.

---

[52] David Allan, Jack Kusters, and Robin Giffard, "Civil GPS Timing Applications," pp. 26-27.

[53] Candidate improvements are discussed in Chapter 4.

The presence of SA, despite the fact that it does not degrade timing accuracy to less than currently acceptable levels, is considered to be another limitation on overall system reliability. The telecommunications industry believes that GPS, being "subject to failures or deliberate denial of signal", cannot and should not be used without being backed up by other technologies able to provide the same information.[54] In the future, it is also likely that accuracies in the range of 50 to 100 nanoseconds will be required for some telecommunications applications. It will be difficult for direct GPS timing to meet this requirement, even without the presence of SA.

Findings

GPS currently meets all accuracy requirements for both GPS time transfer and time synchronization using direct GPS time.

Many telecommunications companies are still hesitant to utilize GPS because of concerns about system reliability and the presence of SA.

Future accuracy requirements for both time synchronization and time transfer will be difficult to achieve with the current capabilities of GPS.

SPACECRAFT USES OF GPS

The application of GPS to spacecraft navigation and control has the potential to provide significant savings in spacecraft costs and mission operations and is being introduced into spacecraft systems today in both government and commercial programs. The feasibility of using GPS for satellite navigation was first demonstrated in 1982 by a receiver placed aboard Landsat 4.[55] Since then a number of additional missions and satellites have utilized GPS, including the Topex/Poseidon satellite launched in 1992. Other spacecraft programs have flown GPS, but it has been used primarily in an experimental mode.

GPS receivers also have been used experimentally for launch vehicle applications. The experimental Ballistic Missile Defense Organization/McDonnell Douglas Delta Clipper (DC-X) utilized a GPS receiver integrated with an inertial navigation unit and flight control avionics during its flight testing. The system has reportedly cut the rocket's development

---

[54] GPS and Global Telecommunications: A Summary Briefing Prepared for the National Research Council Committee on the Future of the Global Positioning System, p. 8.

[55] H. Heuberger and L. Church, "Landsat-4 Global Positioning System Navigation Results," (Presentation to the American Astronautical Society/American Association of Aeronautics and Astronautics (AAS/AIAA) Astrodynamics Conference, AAS 83-363, August 1983).

time and contributed to its success.[56] An integrated GPS/inertial navigation unit is also being test flown on the Orbital Science Corporation's Pegasus launch vehicle. The company hopes that an operational version of the unit will one day improve the vehicle's en route navigation and orbital injection accuracy.

Current and Future Applications and Requirements

GPS is currently being tested or used for several spacecraft applications, including orbit determination, attitude determination, launch and reentry vehicle positioning and trajectory determination, and time synchronization. Precise time synchronization, which is required by many spacecraft, such as telecommunications satellites, to an accuracy of 100 nanoseconds was discussed in some detail in the previous section, but the remaining applications are discussed below.

Orbit Determination

The use of GPS for real-time determination of orbital parameters provides an economical means of determining a spacecraft's orbit very accurately. A properly designed, space-qualified GPS receiver can replace several conventional orbital positioning spacecraft sensors, reducing both weight and cost, and in some cases relieving the requirement for worldwide, ground-based stations to track orbital positions. In addition, the orbital parameters determined with GPS can in some cases be input to an on-board control computer and propulsion system to provide autonomous station keeping. This would alleviate or reduce the need for mission operations personnel to control a spacecraft's orbital position from the ground.

In general the requirements for real-time orbit determination are not very stringent, ranging from about 50 meters to several kilometers. Although these requirements are quite lax, the same is not true for post-flight or post-processed solution accuracies. Many spacecraft, in particular those used for scientific missions, require very precise knowledge of where the satellite was when scientific data were being collected. The desire to achieve ± 1 centimeter orbit determination accuracy for the Topex/Poseidon spacecraft, as discussed in the Earth Science section of this chapter, provides an excellent example. In order to achieve this level of accuracy, GPS measurements from the spacecraft are processed together with GPS data from a worldwide network of ground stations and an extensive set of dynamic models. Future science missions are likely to push this requirement even further towards the millimeter level.

[56] "Delta Clipper Contractors Tout Components' Success," *Space News*, 27 September – 3 October 1993, p. 17. The DC-X is a one-third-scale sub-orbital, single-stage-to-orbit (SSTO) technology demonstrator developed with funding from the Ballistic Missile Defense Organization (BMDO).

Attitude Determination

In the last several years, several manufacturers of GPS receivers have started collaborating with spacecraft developers to design GPS receivers for use as attitude sensors on board spacecraft. On-board attitude determination is a requirement for virtually every modern spacecraft, and most also require an automatic attitude control system. The traditional suite of sensors used for attitude determination range from relatively low-cost magnetometers and horizon sensors to precise gyroscopes, sun sensors, and star trackers. GPS may provide a cost-effective complement or even alternative to many of these existing systems.

GPS attitude determination is accomplished by observing the carrier phase of an incoming GPS signal at two or more antennas on board the spacecraft. The difference in phase between the antennas can be related to the vehicle orientation and the rate of change of these phase observations is an indication of the attitude rate of change. The accuracy of GPS for this application is limited by multipath, the phase noise in the receiver, the separation of the antennas, and the stability of the structure supporting the antennas. With the best current technology, accuracies as good as 0.1 degrees (2σ) can be expected.

The accuracy requirements for satellite attitude determination range from 5 degrees for some simple spacecraft to well below 3×10^{-6} degrees (0.01 arc seconds) for a spacecraft like the Hubble Telescope. At this stage GPS cannot replace the high performance of star trackers for this ultimate precision, but may provide a cost-effective alternative for many mission requirements.

Launch and Re-entry Vehicle Guidance

GPS also has applications to space launch vehicles as a sensor in the vehicle's navigation system and for providing positioning information to ground controllers for range safety purposes. As previously mentioned, an integrated GPS/inertial navigation system has been tested on the experimental BMDO/McDonnell Douglas Delta Clipper (DC-X), and on Orbital Science Corporation's Pegasus launch vehicle. In addition, an experimental space re-entry vehicle called the Spacewedge, designed for re-entry rather than launch, is demonstrating the ability to make an automatic precision landing using a parafoil and a commercial GPS receiver. A full-scale space vehicle, either piloted or unpiloted, may one day use GPS-based technology for emergency crew return or cargo return from Earth's orbit.[57] Accuracy requirements have not been provided for these experimental applications.

Most range safety tracking for launch vehicles currently is conducted using a rather elaborate and expensive system consisting of ground tracking radars and associated equipment. According to a previously published NRC study, it is conceivable that pending

---

[57] Spacewedge, known formally as the "spacecraft autoland gliding parachute experiment," has been developed by NASA's Dryden Flight Research Facility, Edwards AFB, California for under $100,000 annually. J. R. Asker, "Space Autoland System Shows GPS' Wide Uses," *Aviation Week & Space Technology*, 18 October 1993, pp. 54-55.

further study by range safety experts, GPS-derived trajectory data could be used as a more cost-effective alternative.[58] The DOD has been considering the use of GPS as the primary time and space position information source for the national ranges ever since the Range Applications Joint Program Office was established approximately 5 years ago, and the Navy has been utilizing GPS trajectory data for Trident missile testing since the early 1980s.[59] Accuracy requirements for GPS range safety applications are very mission specific, and have not been generalized.

GPS also could be used to improve range safety by sending flight termination commands to missiles and launch vehicles carrying GPS receivers. This could be accomplished using a DGPS datalink or a pseudolite located at the range or, as suggested by one expert in range safety, by using some spare data bits available in the GPS navigation message itself.[60] Current flight termination telecommands, which are used to initiate self-destruction, are broadcast in the UHF frequency band. This band is very susceptible to spoofing, jamming, and interference. Integrating a telecommand with other GPS and DGPS equipment and datalinks already under development for time and position range applications could provide a more secure and cost effective means of initiating a flight termination when it is necessary.

A consolidated list of available GPS requirements for spacecraft applications is provided in Table 2-10.

Table 2-10 Requirements for GPS Spacecraft Applications[a]

	Application	Accuracy
Satellites	Orbit Determination (Real Time)	50 m (2 drms)
	Orbit Determination (Post-Process)	± 0.001 m (2 drms)
	Attitude Determination	5 degrees to 3×10^{-6} degrees (2σ)[b]
Launch Vehicles	Launch Trajectory and Position Determination	Mission Specific

a. Accuracy is currently the only specified requirement for GPS spacecraft applications. The values in this table were derived by the committee from input received during the study.

b. Accuracy as good as 3×10^{-6} degrees (2σ) is currently available only from star trackers. GPS is currently capable of 0.1 degree (2σ) attitude determination accuracy, which is suitable for most spacecraft missions.

---

[58] NRC, *Technology For Small Spacecraft*, Aeronautics and Space Engineering Board, National Research Council (Washington, D.C.: National Academy Press, 1994), p. 16.

[59] Source of Information: Personal conversation with Daniel F. Alves, Jr. of Alpha Instrumentation/Information Management, AI²M, Santa Maria, California, 20 February 1994.

[60] Daniel F. Alves, Jr., *Global Positioning System Telecommand Link*, U.S. Patent number 5,153,598, 6 October 1992.

Challenges to Full GPS Utilization

Orbit Determination and Orbital Positioning

The GPS SPS currently delivers sufficient accuracy for real-time orbit determination requirements. The well-known orbital dynamics of most spacecraft allow filtering of the data, which helps to mitigate the effects of SA. A-S does not have a significant effect on real-time orbit determination because most satellites are above the densest part of the ionosphere and can probably ignore the ionospheric delay contribution.

For precise post-flight requirements SA does not pose a problem, but the presence of A-S reduces the orbit determination accuracy of spacecraft missions which rely on codeless L_2 measurements. Topex/Poseidon, for example, relies on the use of dual-frequency data, but does not carry a receiver capable of tracking the Y-code. The processing of raw data from a large network of stations will still be required, however, even if the basic GPS accuracy is improved. Nevertheless, all improvements to the basic system will aid in the search for the last centimeter or millimeter of precision.

Attitude Determination

Because GPS attitude determination techniques use differential carrier-phase measurements, SA has little or no effect on the accuracy achievable. A-S may have some effect in that it prevents the use of differential P-Code measurements for coarse attitude determination and makes the use of dual frequency differential carrier-phase measurements more difficult. As mentioned previously, however, the accuracy of GPS for this application is limited primarily by design parameters related to receiver electronics and antenna structure.

Signal Visibility

Satellites in orbit near or above the GPS constellation are only able to track GPS signals that pass beyond the limb of the Earth. On the current Block II/IIA satellites there is sufficient antenna beamwidth to allow orbit determination to be performed at geosynchronous altitudes using GPS and a significant amount of dynamic modeling.[61] The Block IIR and IIF satellites, however, may not have the same antenna beamwidth, and the L-band signals broadcast from these antennas may no longer pass beyond the limb of the Earth. This could eliminate the ability of a geosynchronous satellite to receive GPS signals, precluding a potentially important GPS application.

---

[61] S. C. Wu et al., "GPS-Based Precise Tracking of Earth Satellites from Very Low to Geosynchronous Orbits," in *Proceedings of the National Telesystems Conference* (Ashburn, Virginia, May 1992), pp.4-1 to 4-8.

Findings

The presence of SA has little or no effect on the ability to use GPS for orbit determination, but A-S limits the performance of orbit determination for spacecraft that rely on dual-frequency measurements, such as Topex/Poseidon.

SA has no effect on the accuracy of GPS attitude determination methods for spacecraft. A-S may place some limitations on achievable accuracy, but so do design parameters related to receiver electronics and antenna structure.

The ability to use GPS for orbit determination on board geosynchronous satellites will be lost if the Block IIR and IIF spacecraft are built with narrower beamwidth antennas than the Block II/IIA satellites.

SUMMARY

Although this chapter does not represent a complete list of all GPS applications and their requirements, it should be clear from its content that the Global Positioning System has become an integral part of our nation's technical infrastructure, which contributes to our security, economy, and overall quality of life. Indeed, a fully exhaustive list of GPS applications may be impossible to compile, for as soon as it was completed, dozens of new and innovative applications, such as navigation systems for the visually impaired, would be developed that exploit GPS to the limits of its technological capability. Although requirements for currently undiscovered applications such as this one cannot be quantified, a strong case can be made for not only maintaining the basic system's operational capability but also for continuously improving it in order to meet the increasingly demanding requirements of a multitude of military and civilian users who rely on GPS on a routine basis.

The tables included in this summary represent a compilation of the GPS applications that have been discussed in this chapter. Military applications with accuracy requirements currently unmet by the PPS are included in Table 2-11, and civil applications are grouped according to their accuracy requirements in tables 2-12 though 2-16. As these tables and the preceding discussions in this chapter clearly illustrate, the civilian applications that currently require augmentation or enhancement of the GPS SPS far outweigh those that do not. Most integrity and availability requirements for civilian applications are also unmet by the GPS SPS and are highlighted in the tables through the use of grey shading. Candidate technical improvements and modifications to the basic GPS that would enhance its functionality and make it more capable of meeting the requirements of both civilian and military users are discussed in the next two chapters.

Table 2-11 Summary of Military Applications with Accuracy Requirements Unmet by the GPS PPS as Currently Specified[a]

	Application	Accuracy	Integrity	
			1 minus P_{HE} times P_{MD}	Time to Alarm
Aviation	Non-precision Sea Approach/Landings	12.0 m (2 drms)	0.999	10 s
	Precision Approach/Landings Unprepared Surface	12.5 m (2 drms)	0.999	6 s
	Precision Sea Approach/Landings	0.6 m (2 drms)	0.999	6 s
Mine Warfare	Anti-mine Countermeasures	< 5.0 m CEP	Not specified	Not specified
Special Warfare	Combat Swimming	1.0 m CEP	Not specified	Not specified
	Land Warfare & Insertion/Extraction	1.0 m CEP	Not specified	Not specified
Amphibious Warfare	Artillery & Reconnaissance	< 6.0 m CEP	Not specified	Not specified
	Precision-guided Munitions	3.0 m CEP	Not specified	Not specified

a.   References and/or additional notes for each of the requirements listed in this table can be found by referring to previous tables (2-1 through 2-10) included in this chapter.

Table 2-12 Summary of Civilian Applications with Accuracy Requirements of 100 Meters or Greater (currently achievable with the basic GPS SPS)[a]

	Application	Accuracy (2 drms)	Integrity		Availability
			1 minus P_{HE} times P_{MD}	Time to Alarm	
Aviation	En route Oceanic	23 km	Not specified	30 s	99.977%
	En route through Non-precision Approach/Landings	100 m	$1 - 1 \times 10^{-7}$ per hour	8 s	99.999%
	Domestic Automatic Dependent Surveillance (ADS)	200 m	Not specified	Not specified	99.999%
Maritime	Oceanic Navigation	1800 to 3700 m	Not specified	Not specified	99.0%
	Coastal Navigation	460 m	Not specified	Not specified	99.7%

a.   References and/or additional notes for each of the requirements listed in this table can be found by referring to previous tables (2-1 through 2-10) included in this chapter.

Table 2-13 Summary of Civilian Accuracy Requirements Between 25 and 100 Meters[a]

	Application	Accuracy (2 drms)	Integrity		Availability
			1 minus P_{HE} times P_{MD}	Time to Alarm	
ITS and Vehicle Navigation/Position-Location	Fleet Management (AVL/AVI)	25 to 1500 m	Not specified	1 to 15 s	99.7%
	Emergency Response	75 to 100 m	Not specified	1 to 15 s	99.7%
	Vehicle Command and Control	30 to 50 m	Not specified	1 to 15 s	99.7%
	Accident Data Collection	30 m	Not speciified	1 to 15 s	99.7%
Spacecraft (Satellites)	Orbit Determination (real time)	50 m	Not specified	Not specified	Not specified

a.   References and/or additional notes for each of the requirements listed in this table can be found by referring to previous tables (2-1 through 2-10) included in this chapter.

Table 2-14 Summary of Civilian Accuracy Requirements Between 10 and 25 Meters[a]

	Application	Accuracy (2 drms)	Integrity		Availability
			1 minus P_{HE} times P_{MD}	Time to Alarm	
Aviation	TCAS	14.4 m	Not specified	Not specified	Several Days
	Surface Surveillance	24.0 m	Not specified	Not specified	99.87%
Maritime	Recreational Boating	10.0 m	Not specified	Not specified	99.9%
	Vessel Traffic Services	10.0 m	Not specified	Not specified	99.9%
ITS	Infrastructure Management	10.0 m	Not specified	1 to 15 s	99.7%
Search & rescue	Location Determination	10.0 m	Not specified	minutes	99.0%
Oceanography	Real-time Navigation and Positioning	10.0 to 30.0 m	Not specified	Not specified	Not specified

a.   References and/or additional notes for each of the requirements listed in this table can be found by referring to previous tables (2-1 through 2-10) included in this chapter.

Table 2-15 Summary of Civilian Accuracy Requirements Between 1 and 10 Meters[a]

	Application	Accuracy (2 drms)	Integrity — 1 minus P_{HE} times P_{MD}	Integrity — Time to Alarm	Availability
Aviation	CAT I Approach/Landing	7.6 m	$1 - 4 \times 10^{-8}$ per approach	5.2 s	99.9%
	CAT II Approach/Landing	1.7 m (vertical)	$1 - 0.5 \times 10^{-9}$ per approach	2.0 s	Not specified
Maritime	Harbor/Harbor Approach Navigation	8.0 to 20.0 m	Not specified	6 to 10 s	99.7%
	Inland Waterway Navigation	3.0 m	Not specified	6 to 10 s	Not specified
Railroad	Train Control	1.0 m	Not specified	5 s	99.7%
ITS and Vehicle Navigation/ Position-Location	Highway Navigation and Guidance	5.0 to 20.0 m	Not specified	1 to 15 s	99.7%
	Mayday/Incident Alert	5.0 to 30.0 m	Not specified	1 to 15 s	99.7%
	Automated Bus/Rail-Stop Annunciation	5.0 to 30.0 m	Not specified	1 to 15 s	99.7%
	Collision Avoidance, Control	1.0 m	Not specified	1 to 15 s	99.7%
	Collision Avoidance, Hazardous Situation	5.0 m	Not specified	1 to 15 s	99.7%
Hazmat Transport	Vehicle or Cargo Location	5.0 m	Not specified	1 s	99.7%
Land Recreation	Off-road Vehicles, Hikers, Back-country Skiers, etc.	5.0 m	Not specified	Minutes	99.0%
Earth Science	Airborne Geophysics	3.0 m (vertical)	Not specified	Minutes	Not specified
Mapping/ Surveying	Geographic Information Systems (GIS)	1.0 to 10.0 m	Moderate	Not specified	98%

a.   References and/or additional notes for each of the requirements listed in this table can be found by referring to previous tables (2-1 through 2-10) included in this chapter.

Table 2-16 Summary of Submeter Civilian Accuracy Requirements[a]

	Application	Accuracy (2 drms)	Integrity		Availability
			1 minus P_{HE} times P_{MD}	Time to Alarm	
Aviation	CAT III Approach/Landing	0.6 to 1.2 m (vertical)	$1 - 0.5 \times 10^{-9}$ per approach	2.0 s	Not specified
Precision Farming	Automatic Vehicle Guidance	0.05 m	Not specified	5.0 s	99.7%
Mapping/ Surveying/ Geodesy	Photogrammetry	0.02 to 0.05 m	Not specified	Minutes	98.0%
	Remote Sensing	0.1 to 20.0 m	Not specified	Not specified	98.0%
	Geodesy	0.01 to 0.05 m	Not specified	Hours	98.0%
	Mapping	0.1 to 10.0 m	Not specified	Hours	98.0%
	Surveying	0.01 to 10.0 m	Not specified	Hours	98.0%
Earth Science	Oceanography (ocean circulation determination)	0.01 m	Not specified	Hours	Not specified
	Geodynamics	0.001 m + $10^{9} \times$ baseline length	Not specified	Hours	Not specified
Spacecraft (satellites)	Orbit Determination (post-process)	$\pm$ 0.001 m	Not specified	Not specified	Not specified
	Attitude Determination	3×10^{-6} degrees (0.01 arc second), 2σ	Not specified	Not specified	Not specified

a.   References and/or additional notes for each of the requirements listed in this table can be found by referring to previous tables (2-1 through 2-10) included in this chapter.

3

Performance Improvements to the Existing GPS Configuration

INTRODUCTION

As pointed out in the previous section, civil users of the GPS have accommodated themselves to the currently available SPS (Standard Positioning Service) in attempting to meet their individual performance requirements, and a number of innovative uses of GPS have been demonstrated with the existing system. An even more capable system would likely result in a larger number of applications. Improved accuracy, integrity, availability, and reliability of the signal could provide improved results at significantly lower cost. For example, if the stand-alone GPS could provide an accuracy approaching 5 meters (2 drms), the need for many of the existing or planned differential systems could be avoided.

In accordance with the committee's statement of task, this chapter will recommend a sequence of enhancements to the GPS that will serve to improve the accuracy of the system for civilian, commercial, and military users. After a discussion of the current performance achievable from the basic GPS, the subsequent sections address specific accuracy improvements focused on enhancing civilian, commercial, and military use of the system. Many of the suggested improvements also will have benefits other than better accuracy, such as increased integrity, improved availability, and enhanced resistance to RF (radio frequency) interference. These improved characteristics are discussed where appropriate. The final section of this chapter presents an overall strategy for implementing the recommended improvements. As noted throughout the text, some of the improvements are meant to be applied to the current GPS satellite constellation and others to the Block IIR and Block IIF constellations. When available, the approximate cost of each improvement also is given.

CURRENT GPS PERFORMANCE

Accuracy

As can be seen from Table 3-1, the contributors to civilian SPS signal accuracy errors are SA (Selective Availability), the atmospheric error, the clock and ephemeris errors, the receiver noise error, and the multipath error. For the military PPS (Precise Positioning Service) signal, the largest error contributors are the clock and ephemeris errors, the receiver noise, and multipath errors, since the PPS signal is not degraded by SA. The ionospheric error for the PPS signal is small relative to that for the SPS signal since the military has access to both the L_1 and L_2 frequencies and can correct for the ionospheric error.

Table 3-1 Observed GPS Positioning Errors with Typical SPS and PPS Receivers[a]

Error Source	Typical Range Error Magnitude (meters, 1σ)	
	SPS with II/IIA satellites	PPS with II/IIA satellites
Selective Availability[b]	24.0	0.0

Atmospheric Error
  Ionospheric[c]
  Tropospheric[d]
	7.0
	0.7
		0.01
		0.7

Clock and Ephemeris Error[e]	3.6	3.6
Receiver Noise[f]	1.5	0.6
Multipath[g]	1.2	1.8
Total User Equivalent Range Error (UERE)	25.3	4.1
Typical Horizontal DOP (HDOP)[h]	2.0	2.0
Total Stand-Alone Horizontal Accuracy, 2 drms[i]	101.2	16.4

a.   It is assumed here that a "typical" SPS and PPS receiver has a four-satellite position solution.

b.   J. F. Zumberge and W. I. Bertiger, "Ephemeris and Clock Navigation Message Accuracy in the Global Positioning System," Vol. I, Chap. 16. Edited by B. W. Parkinson, J. J. Spilker, P. Axelrad, and P. Enge (To be published by AIAA, in press 1995). This error is manifested as increased clock and ephemeris errors when SA is on.

c.   For the SPS signal, the ionospheric content is quite variable, with large diurnal variations, and large variations over the 11-year solar cycle. Depending on the Total Electron Content (TEC), a delay at L_1 ranging from less than 1 meter to 70 meters can result. A typical SPS receiver has an algorithm that can remove about 50 percent of the ionospheric error, leading to an error ranging from less than 1 meter to 35 meters. For the above table, an error of 7 meters was used, which is typical for a daytime mid-latitude ionospheric error near the maximum of the 11-year solar cycle, after correction by the standard algorithm. Because the ionospheric error is not independent between satellites, it should not strictly be considered a range error to be multiplied by HDOP (Horizontal Dilution of Precision). When the ionospheric content is uniform above the receiver, such as during the pre-sunrise morning, it contributes little to horizontal error, but maps into errors in the vertical position and receiver clock. When there are significant gradients in the ionospheric content, however, such as exist at local dawn and dusk, errors are induced into the horizontal position. Therefore, the use of 7 meters for a range error, which is multiplied by HDOP, is a somewhat conservative choice. For the PPS signal, the ionospheric error is removed by a linear combination of the L_1 and L_2 observables. This correction leaves residual ionospheric error of 1 centimeter or less.

d. For a typical SPS or PPS receiver, software models correct for all but around 0.7 meters (1σ) of the tropospheric error. The tropospheric error is even more highly correlated than the ionospheric error, due to its uniform distribution. The errors introduced by the troposphere normally map into the vertical position and receiver clock errors. As for the ionospheric error, the multiplication of this error by HDOP to obtain the horizontal error is a conservative calculation.

e. This value is based on observed data as noted in "Ephemeris and Clock Navigation Message Accuracy in the Global Positioning System." (See note a above). The combined clock and ephemeris error does not contain SA epsilon error in the broadcast ephemeris nor the SA dither error in the broadcast time.

f. For a SPS receiver, the receiver noise for independent 1-second measurements can actually range from around 0.25 to 2.0 meters, depending on its design. For a PPS receiver, the single-frequency pseudorange noise error is less because the ten times faster Y-code chip rate overcomes the 3 dB to 6 dB signal-to-noise ratio penalty relative to the C/A code. In forming the linear combination required to removed the ionospheric error, Y-code corrected = $2.55(Y)L_1 - 1.55(Y)L_2$, the noise error of the Y-code is effectively multiplied by the root sum square of 2.55 and 1.55, which is approximately 3. (A single-frequency PPS receiver like the Plugger would have a receiver range error smaller by a factor of three, but at the cost of retaining a 7-meter error due to the ionosphere). The PPS receiver noise error can range from 0.1 to 0.8 meters (1σ), for independent 1-second measurements.

g. For a SPS receiver, multipath can typically range from 0.4 to 5 meters (1σ), depending on the antenna, antenna surroundings, and receiver design. For a PPS receiver, the single-frequency multipath error is somewhat less, typically by a factor of 0.5, because of the faster chip rate. In forming the linear combination required to remove the ionospheric error, Y-code corrected = $2.55(Y)L_1 - 1.55(Y)L_2$, the Y-code multipath error is effectively multiplied by the root sum square of 2.55 and 1.55, which is approximately 3. This explains why the PPS multipath error exceeds the SPS multipath error. (A single-frequency PPS receiver like the Plugger would have a multipath error smaller by a factor of three, but at the cost of retaining a 7-meter error due to the ionosphere). The PPS multipath error can range from 0.3 to 2 meters.

h. HDOP can vary depending on the geometry of the satellites. For a typical SPS or PPS receiver, the geometric strength of a four-satellite solution is limited, so a conservative HDOP of 2.0 was used.

i. These values are based on observation and differ from the accuracy values specified by the DOD (Department of Defense), shown in Figure C-7, Appendix C.

Specific technical modifications to GPS to reduce the errors discussed above and improve the accuracy for both the military and civilian communities are discussed in detail below. As explained in the table notes above, the exact numbers in the tables can vary. If all of the recommendations are implemented, the committee believes that the stand-alone horizontal GPS accuracy will approach 5 meters (2 drms).

Greater stand-alone accuracy could take the place of differential GPS systems for some users who require accuracies of a few meters (2 drms). For example, greater stand-alone GPS accuracy would allow many vehicle positioning and navigation requirements to be met without the use of DGPS. To use a military example, precision weapons, such as missiles and smart bombs that have been equipped with GPS, presently require expensive terminal guidance packages or access to differential corrections to meet accuracy requirements of a few meters (CEP). In order to take advantage of GPS accuracy, accurate knowledge of the target location is essential. Various target-determination techniques are being developed, but until they are available, munitions delivery, even with GPS, will still require highly accurate terminal guidance systems. Using an enhanced GPS with greater accuracy for guidance would provide two levels of benefit. First, the requirements on an inertial navigation device can be relaxed because more accurate GPS determination of position and velocity will be possible. Second, under some conditions where jamming is not likely, GPS could be used to provide a very economical and accurate stand-alone munitions guidance system.

Integrity and Availability

In general, improving the system ranging accuracy also improves integrity and availability. As noted in Appendix C, availability is the percentage of time that a user's positioning errors lie within a specified accuracy. If the ranging errors are decreased, then positioning errors will remain inside the accuracy bounds for higher DOP (Dilution of Precision) values. As a consequence, the amount of time that the system is available increases, especially in the presence of satellite outages. Improvements in availability for the SPS as ranging accuracy improves will be shown throughout this chapter by comparing availability values for Chicago, Illinois.[1]

Integrity checking algorithms also benefit from improved ranging accuracy. Most integrity algorithms, such as RAIM (Receiver Autonomous Integrity Monitoring), are based on consistency checks among redundant sets of measurements.[2] Poor consistency indicates the possibility of a position solution error exceeding the protection limit. However, when the range errors are large, the consistency checks are not reliable except under very favorable satellites geometries. With improved range measurement accuracy, consistency can be reliably measured even under poor user-satellite geometries.

SELECTIVE AVAILABILITY AND ANTI-SPOOFING

The GPS was designed to provide our military forces with an advantage when engaged with other military forces, while still providing a reasonable positioning service to

---

[1] Chicago, Illinois, was randomly chosen by the MITRE Corporation, which determined the availability values presented in this chapter using a GPS availability model developed for the FAA (Federal Aviation Administration). The analytical model accounts for individual satellite short-term and long-term failures and restorations for the 24-satellite constellation and assumes a conservative serial restoration strategy (that is, only one satellite can be replaced at a time). The GPS receiver was assumed to have an elevation mask angle of 5 degrees.

[2] RAIM, which utilizes receiver software algorithms to detect unreliable satellites or position solutions, is defined in Appendix C.

the civil community. Two features were incorporated into the GPS to provide this advantage. The first, SA, degrades the GPS signal so that the unencrypted signal on L_1 will provide a stand-alone horizontal accuracy of 100 meters (2 drms).[3] The second, A-S (Anti-Spoofing), provides encryption of the P-code on L_1 and L_2 in order to deny the signal to the adversary and to increase resistance to spoofing.

Selective Availability

Currently, the full accuracy of GPS is denied to stand-alone non-PPS users of GPS for both navigation and time transfer through the implementation of SA. SA comprises two functions: (1) fluctuation of the GPS satellite clock frequency, known as dither, and (2) transmission of incorrect ephemeris parameters in the navigation message, termed epsilon. SA affects all GPS observables, which include the C/A-code and P-code pseudorange measurements and the L_1 and L_2 carrier phase measurements. SA is discussed in greater detail in Appendix C. The DOD has stated that the degradation produced by SA will be limited to a value that maintains the 100-meter (2 drms) specified stand-alone horizontal accuracy of the SPS. Furthermore, at a recent meeting of the DOD/DOT (Department of Transportation) Signal Specification Issues Technical and Policy Groups, additional specifications were discussed and agreed upon for limits on the individual satellite range rate and acceleration errors, shown in Table 3-2.[4]

Table 3-2 SA Errors from DOD/DOT Signal Specification Issues Technical Group

Type of Error	Specification
Range Rate Bound	Not to exceed 2 m/s
Range Acceleration Bound	Not to exceed 19 mm/s^2
Range Acceleration	8 mm/s^2 (2σ)

Under special circumstances, the level of SA errors can be set to zero or increased to a larger value, but only by the National Command Authority. For example, SA was set at a very low level during the Persian Gulf War and during the initial occupation of Haiti

---

[3] The Under Secretary of Defense for Research and Engineering officially established the 100-meter (2 drms) accuracy level for the SPS on June 28, 1983. This policy is reiterated in each biannual publication of the *Federal Radionavigation Plan.*

[4] *Report of the DOD/DOT Signal Specification Issues Technical Group to the Policy Group,* Washington, D.C., 13 December 1994.

(it was then set back to 100 meters, 2 drms), because of the lack of PPS equipment fielded by the U.S. military at those times.

PPS receivers are able to completely remove the effects of both SA dither and epsilon from their observations through the use of a security module. SPS receivers can eliminate the effects of SA through the use of local or wide-area DGPS broadcasts of differential corrections. DGPS reference stations typically broadcast observed range and range-rate errors. The level of SA-induced range acceleration determines the rate at which the corrections must be updated to keep the user error within acceptable bounds. Satellite position errors produced by the epsilon technique will decorrelate as the separation between the reference station and the user increases. Wide-area DGPS will provide orbit corrections for each satellite to compensate for this effect.

Military Utility of SA

The DOD has stated that SA is an important security feature because it prevents a potential enemy from directly obtaining positioning and navigation accuracy of approximately 12.5 meters CEP (30 meters, 2 drms) from the C/A-code. Since the military has access to a specified accuracy of 8 meters CEP (21 meters, 2 drms), they believe U.S. forces have a distinct strategic and tactical advantage. With SA at its current level, a potential enemy has access only to the 42-meter CEP (100 meters, 2 drms) accuracy available from the SPS. The DOD believes that obtaining accuracies better than 42 meters CEP requires a substantial amount of effort. DOD representatives have expressed their belief that our adversaries are much more likely to exploit the GPS C/A-code, rather than DGPS, because its use requires less effort and technical sophistication than is required to use DGPS. In addition, some DOD representatives contend that local-area DGPS broadcasts do not diminish the military advantage of SA because they could be rendered inoperative, if warranted, through detection and destruction or by jamming.

It is the opinion of the NRC committee however, that meter-level accuracies are readily obtainable, even in the presence of SA set at its current level or even at higher levels. As shown in Figure 3-1, several DGPS systems, operated by both commercial and government entities, routinely provide position accuracies approaching 1 meter (2 drms) in the United States and in most of the populated areas of the world. Further information on commercially available systems is provided in Appendix C. Even within the U.S. government, civilian agencies such as the Federal Aviation Administration, the Coast Guard, and the Army Corp of Engineers are planning to operate systems that will, in combination, cover the entire United States and beyond, as shown in Figure 3-2. Furthermore, if the full GLONASS constellation is completed in 1995 as currently planned, this system also will provide properly equipped users with an additional source of highly accurate positioning data, as shown in Figure 3-3.[5]

---

[5] Unlike GPS, GLONASS does not deny accuracy to some users through the use of SA or a similar technique.

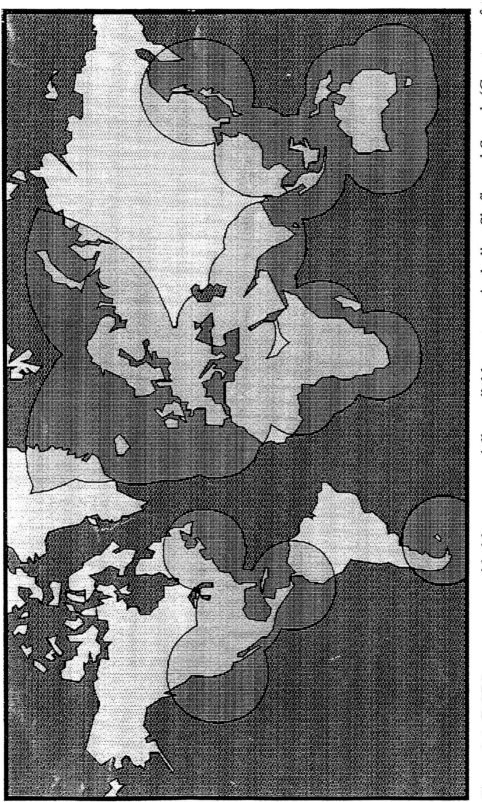

Figure 3-1 DGPS coverage provided by commercially available systems, including Skyfix and Sercel. (Courtesy of the National Air Intelligence Center)

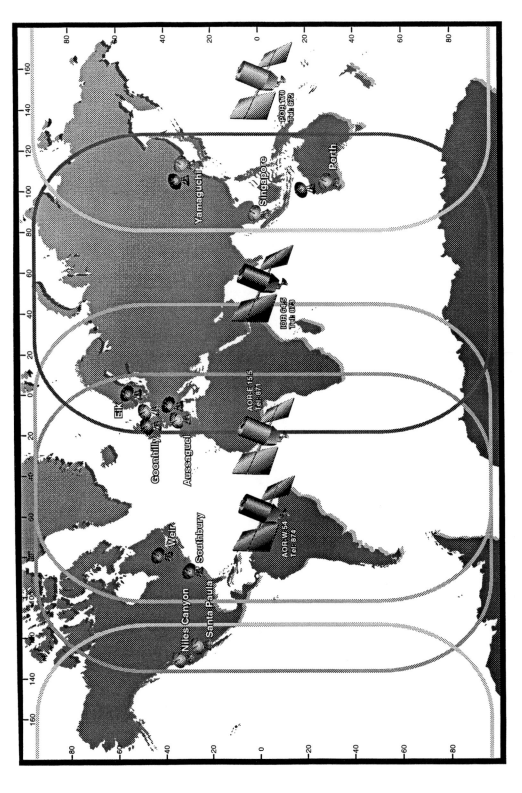

Figure 3-2 DGPS coverage provided by the planned FAA WAAS (Wide-Area Augmentation System). Source: Innovative Solutions International, Inc., presentation at the National Technical Meeting of the Institute of Navigation Meeting, Anaheim, California, January 1995.

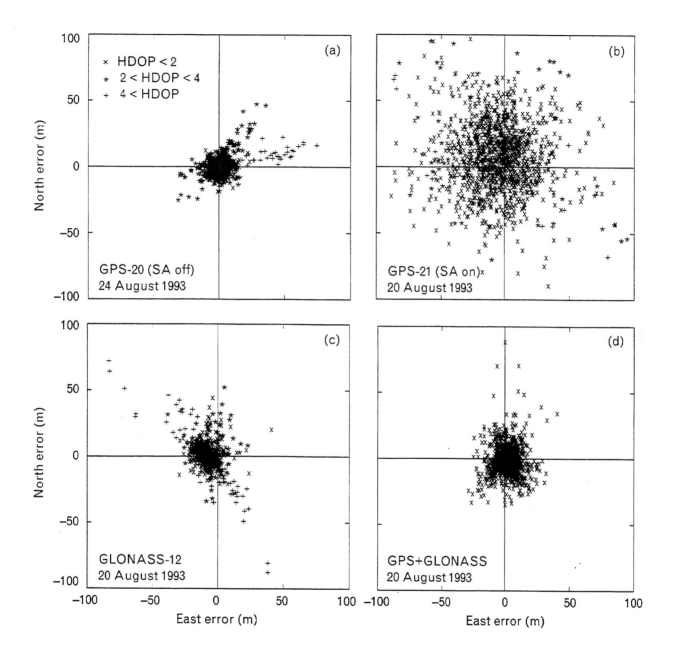

Figure 3-3 Position estimates from GPS and GLONASS obtained from measurement snapshots taken 1 minute apart over an entire day. Position from (a) GPS with SA off, (b) GPS with SA on, (c) GLONASS, and (d) GPS plus GLONASS. (Courtesy of MIT Lincoln Laboratory)

Even if potential adversaries are not taking advantage of DGPS at this time, the NRC committee believes that it would be prudent for the DOD to recognize the potential capability that currently exists. In addition, the establishment of a low-cost, militarily controlled local-area DGPS network for use by an adversary in a theater of conflict is an even more likely possibility. Local-area differential systems are easy to build or buy and are inexpensive. Furthermore, the NRC committee believes that the detection and elimination of these military local-area DGPS stations, either in wartime or peacetime, would be difficult. Local-area DGPS reference stations are small and can be installed in less than an hour. Signals from such systems are difficult to detect because they can be broadcast at low power and at spread-spectrum frequencies or in rapid on/off cycles, with very short transmission times. Therefore, they are not easy to detect electronically or visually.

The NRC committee expects that any enemy of the United States sophisticated enough to operate GPS-guided weapons will be sophisticated enough to acquire and install local-area differential system or take advantage of an existing commercial system. These systems can have the capability to provide velocity and position corrections to cruise and ballistic missiles with accuracies that are equal to or superior to those available from an undegraded C/A-code. Even if the level of SA is increased, DGPS methods could still be used to provide an enemy with accurate signals. Further, as previously mentioned, if the full GLONASS constellation is completed in 1995 as currently planned, this system also will provide properly equipped users with an additional source of highly accurate positioning data.

The unencrypted C/A-code, which is degraded with SA, still provides our adversaries with an accuracy of 100 meters, 2 drms (42 meter CEP), which would still be more than adequate to deliver chemical, biological, or even explosive weapons, if creating terror in a city is the enemy's objective (see Figure 3-4). Further, any enemy encountered is not likely to share the U.S. military's interest in limiting collateral damage. With SA set at zero, the stand-alone accuracy improves to 30 meters, 2 drms (approximately 13 meters CEP) or better, depending on the solar cycle and user equipment capabilities. While this improvement enhances the ability of an adversary to successfully attack high-value point targets, significant damage also can be inflicted with accuracies of 100 meters, 2 drms. Therefore, in either case (30-meter or 100-meter accuracy, 2 drms) the NRC committee believes that the risk is sufficiently high to justify denial of the L_1 signal by jamming. The jamming strategy has the additional benefit of denying an adversary all radionavigation capability including the even more accurate DGPS threat.

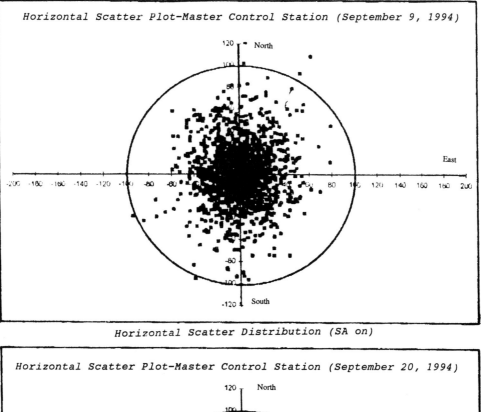

Horizontal Scatter Distribution (SA on)

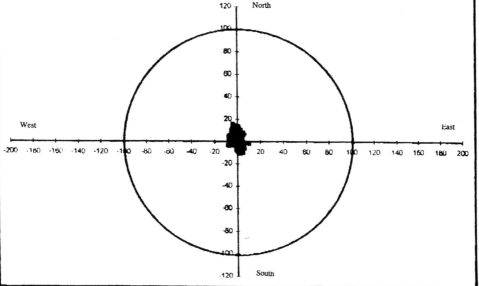

Horizontal Scatter Distribution (SA off)

Figure 3-4 Horizontal scatter plot of 42 meters CEP (100 meters, 2 drms) with SA at its current level and horizontal scatter plot of approximately 10 meters CEP (24 meters, 2 drms) without SA. (Figure Courtesy of Mr. Jules McNeff, Office of the Assistant Secretary of Defense, C³I)

The NRC committee strongly believes that preservation of our military advantage with regard to radionavigation systems should focus on electronic *denial* of all useful signals to an opponent, for example, by jamming and spoofing, while improving the ability of civil and friendly military users to employ GPS in a jamming and spoofing environment. Continued effort to deny the accuracy of GPS to all users except the U.S. military via SA appears to be a strategy that ultimately will fail. Thus, the NRC committee recommends that the military employ jamming techniques in a theater of conflict to fully deny an enemy the use of GPS or other radionavigation systems.

The NRC committee believes that the principal shortcoming in a jamming strategy, regardless of the level of SA, is the difficulty military GPS receivers currently have acquiring the Y-code during periods when the C/A-code is unavailable due to jamming of the L_1 signal.[6] The implementation of direct Y-code acquisition capability, as recommended later in this chapter, would provide the optimal solution to this problem. In the interim, various operating disciplines, also discussed in this chapter, can minimize the impact of L_1 C/A-code jamming on the ability to acquire the Y-code. The committee believes that a focused, high-priority effort by the DOD to develop and deploy direct Y-code user equipment, backed by forceful political will from both the legislative and executive branches, can bring about the desired result in a relatively short period of time. The technology for developing direct Y-code receivers is available today.

Impact of SA on GPS User Equipment Manufacturers and U.S. Competitiveness

It has been argued that SA provides a competitive advantage to U.S. manufacturers of GPS and DGPS user equipment, and DGPS service providers. This has apparently been true in the past and to some extent currently. However, the advantage is at best temporary, as indicated by growing foreign competition, especially from Japan. Foreign manufacturers already possess the technology to achieve results equivalent to those of U.S. manufacturers. Within 1 to 2 years, any competitive advantage for U.S. manufacturers will disappear.

One market analysis has shown that if SA is eliminated, the number of GPS and DGPS users in North America is expected to increase substantially. The market for GPS receivers and systems is estimated to be around $64 billion by the year 2004, as compared to $42 billion with SA at its current level.[7]

There is considerable concern within the U.S. civil user community, and even more concern among the international community, regarding the reliability of a navigation system under the control of the U.S. military. Removal of the SA signal degradation is likely to be viewed as a good faith gesture by the civil community and could substantially improve international acceptance and potentially forestall the development of rival satellite navigation systems.

---

[6] The C/A-code is normally used initially to acquire the Y-code.

[7] The analysis by Michael Dyment, Booz•Allen & Hamilton, 1 May 1995, is shown in Appendix E.

Impact of SA on Civil Users

Turning SA to zero, or completely eliminating SA, would have an immediate positive impact on civil GPS users. The following benefits can be expected:

- improved stand-alone navigation, positioning, and timing accuracy;

- improved availability for any given positioning accuracy;

- enhanced ability to perform RAIM;

- reduced data rate requirements for DGPS corrections;

- enable system modifications that further improve accuracy; and

- improved WAAS.

Each of these benefits is discussed further below.

Increased Stand-Alone Navigation, Positioning, and Timing Accuracy. The stand-alone accuracy for SPS users would immediately increase from 100 meters (2 drms) to around 30 meters (2 drms) if SA were turned to zero, as shown in Table 3-3.[8] For many users currently employing DGPS techniques, such as emergency response vehicles, accident data collection, and vehicle command and control, stand-alone horizontal accuracy of approximately 30 meters (2 drms) is sufficient. Currently, DGPS-equipped receivers cost substantially more (several hundred dollars) than a stand-alone receiver. Savings would result from the elimination of the need for a DGPS receiver and electronics to insert the messages to the GPS receiver. Savings also will result from elimination of the user fee imposed by private DGPS providers.

---

[8] Recent measurements with SA off have ranged from 5 meters to 10 meters (2 drms). However, the accuracy without SA greatly depends on the condition of the ionosphere at the time of observation and user equipment capabilities.

Table 3-3 The Effect of Eliminating SA on GPS SPS Stand-Alone Horizontal Accuracy[a]

Error Source	Typical Range Error Magnitude (meters, 1σ)	
	SPS with SA (II/IIA Satellites)	SPS with No SA (II/IIA Satellites)
Selective Availability	24.0	0.0
Atmospheric Delay		
Ionospheric	7.0	7.0
Tropospheric	0.7	0.7
Clock and Ephemeris Error	3.6	3.6
Receiver Noise	1.5	1.5
Multipath	1.2	1.2
Total User Equivalent Range Error (UERE)	25.3	8.1
Typical Horizontal DOP (HDOP)	2.0	2.0
Total Stand-Alone Horizontal Accuracy, 2 drms	101.2	32.5

a.   All footnotes to Table 3-1 also apply to Table 3-4.

Improved Availability. As explained earlier in this chapter, GPS availability is directly related to accuracy. When the stand-alone horizontal accuracy of the system improves to around 30 meters (2 drms), the availability of any accuracy greater than 30 meters will increase. For example, the average observed availability of the 100-meter (2 drms) SPS for a receiver located in Chicago, Illinois is currently 99.2 percent. For the same 100-meter accuracy level with SA removed, the availability would increase to approximately 99.94 percent.[9]

Enhanced Integrity Monitoring. The ability of a receiver to detect invalid GPS pseudorange measurements autonomously also would be greatly enhanced if SA were turned to zero. RAIM is generally possible if six or more satellites are visible and are providing pseudorange accuracies that allow the easy detection of an inaccurate signal. With SA set at its current level, each satellite range may be in error by 25 meters (1σ) or more, as shown in Table 3-3. This makes it difficult to distinguish a failure. Without SA, pseudorange accuracy improves to almost 8 meters (1σ), dramatically improving the ability to isolate specific satellite faults, as well as signal tracking problems within the receiver itself. An analysis of the impact on RAIM with the elimination of SA was conducted for this study by the MITRE Corporation. The improved RAIM capability has been quantified in terms of

---

[9] Based on analysis conducted by the MITRE Corporation for the Memorandum from the MITRE Corporation to the NRC committee, 7 February 1995. For more details, see footnote 1 earlier in this chapter.

the availability of six useable satellites for three phases of aircraft flight. These results are shown in Table 3-4 and discussed further in Appendix F.

Table 3-4 Effect of SA Removal on RAIM Availability for Aviation Applications[a]

Aviation Application		Availability With SA at its Current Level		Availability With SA Turned to Zero	
Phase of Flight	Protection Limit	21 Satellites[b]	24 Satellites[c]	21 Satellites	24 Satellites
En Route	2.0 nautical miles	93.16%	99.89%	96.34%	99.98%
Terminal Area	1.0 nautical miles	89.96%	94.39%	94.39%	99.95%
Non-Precision Approach	0.3 nautical miles	80.89%	98.88%[d]	91.10%	100.00%[d]

a.   This analysis has been made for a single-frequency C/A-code receiver aided by a barometric altimeter (required for aviation supplemental navigation use of GPS) with a visibility mask angle of 5 degrees.

b.   The probability of having 21 satellites operating is assumed to be 98 percent.

c.   The probability of having 24 satellites operating is assumed to be only 70 percent. However, the values in this table reflect the fact that if 24 satellites are fully operational, an incremental improvement in availability exists.

d.   Although these values would intuitively be lower than the 1 nautical mile terminal area protection limit value, availability improves for the 0.3 nautical mile non-precision protection limit because the barometric altimeter inputs provide extra information in this phase of flight.

Reduced Data Rate Requirements for DGPS Corrections. In addition to reduced receiver costs and DGPS provider fees, a stand-alone horizontal positioning accuracy of approximately 30 meters (2 drms) would allow users to avoid the complexity and expense of receiving differential corrections or post-processing their data. Users requiring accuracies from around 1 meter to 30 meters could still use DGPS, but at a much reduced update rate.[10]

---

[10] The required update rates are derived below, assuming 0.2 meters is allotted to the clock portion of the differential correction for SA at its present nominal level and for SA turned to zero. In addition, this analysis is only valid assuming that precise range-rate information is provided in the navigation message. The result is that the update rate is about two orders of magnitude lower when SA is turned to zero. This advantage would be less for lower accuracy requirements. Other requirements may force higher update rates for specific differential users.

Example with SA at current level:
The 1σ SA range acceleration is 0.004 m/s² from Table 3-2. In order to calculate the update rate required for differential corrections, set $0.2 \text{ m} = 0.5(a)(t^2)$, where $a = 0.004$ m/s². Solving for t results in a required update period of $t = 10$ seconds.

Enable System Modifications that Further Improve Accuracy. If SA is turned to zero, then accuracy is limited by ionospheric errors, clock and ephemeris errors, multipath errors, and receiver noise errors, as Table 3-3 illustrates. As discussed later, technical modifications can reduce these errors. However, with SA set at its current level, any modifications to reduce other errors and improve accuracy will be overwhelmed by the degrading effects of SA.

Improved WAAS. When SA dithering of the GPS signals is employed, the DGPS corrections required to circumvent SA accuracy degradation must keep up with the dithering rate. Since WAAS will broadcast its differential corrections as part of the navigation message data carried by a GPS-like L_1 signal, a high-data rate for the differential correction is required, which constrains the flexibility of providing additional information on the navigation message. If SA were eliminated, the data rate requirement could be relaxed and more information, such as GPS integrity information and other safety or air traffic control related information, could be sent to the user. As noted above, integrity also would improve if SA were eliminated, However, even if SA were removed, the FAA's integrity and availability requirements would still not be met with the basic GPS. Some type of augmentation, such as WAAS, would still be required.

Findings and Recommendations

The NRC committee finds that in view of the rapid proliferation of both local and wide-area DGPS systems worldwide and the ease with which local DGPS stations can be deployed, the current effectiveness of SA in deterring precision attack by adversary forces is severely limited and will essentially be ineffective in the near future.

The NRC committee also found that effective countermeasures to adversary use of GPS and DGPS are currently inadequate. The NRC committee believes that future military strategy should focus on electronic *denial* of all useful signals to our enemies, for example, by jamming and spoofing, while improving U.S. military ability to use GPS in a jamming and spoofing environment.

The principal shortcoming in this strategy, regardless of the level of SA, is the difficulty military GPS receivers currently have in acquiring the Y-code during periods when the C/A-code is unavailable due to jamming of the L_1 signal. The implementation of direct Y-code acquisition capability, as recommended later in this chapter, would provide the optimal solution to this problem. Based on information from receiver manufacturers, the committee believes that the technology for developing direct Y-code receivers is available

---

Example with SA turned to zero:

In this case, the error in the differential correction due to the satellite clock does not include any clock dithering, and so is dominated by the satellite oscillator stability, which is $\Delta f/f = 5 \times 10^{-13}$. Using the formula: $0.2 \text{ m} = t(c)(\Delta f)/f$ to calculate the range error, where c is the speed of light $= 3 \times 10^8$ m/sec, gives a required update period of t = 1,333 seconds (22.2 minutes).

today. The committee believes that a focused high priority effort by the DOD to develop and deploy direct Y-code user equipment, backed by forceful political will from both the legislative and executive branches, can bring about the desired result in a relatively short period of time. However, in the interim time before direct Y-code receivers are fielded by the military, various operating disciplines also discussed in this chapter, can minimize the impact of L_1 C/A-code jamming on the ability to acquire the Y-code.

The committee also has taken cognizance of the DOD belief that exploitation of the GPS C/A-code is more likely in the near term than exploitation of DGPS signals. Even if potential adversaries are not taking advantage of DGPS at this time, the NRC committee believes that it would be prudent for the DOD to recognize the potential capability that currently exists.

The NRC committee believes that continued reliance on SA as a means of denying precise GPS position location to all non-military users over a wide area is a strategy that will ultimately fail. In addition, the removal of SA and the subsequent increase in accuracy obtainable by civil and commercial GPS users would have substantial benefits, as previously discussed. If the use of SA is eliminated, the NRC committee also expects that the market for GPS receivers and systems would increase substantially, as discussed further in Appendix E.

The six most important findings of the NRC committee regarding the impact of SA on the various classes of civilian users and on meeting its intended purpose are

(1) The military effectiveness of SA is significantly undermined by the existence and widespread proliferation of DGPS augmentations as well as the potential availability of GLONASS signals.

(2) Turning SA to zero would have an immediate positive impact on civil GPS users. Without SA, the use of DGPS would no longer be necessary for many applications. System modifications that would further improve civilian accuracy also would be possible without SA.

(3) Deactivation of SA would likely be viewed as a good faith gesture by the civil community and could substantially improve international acceptance and potentially forestall the development of rival satellite navigation systems. Without SA, the committee believes that the number of GPS and DGPS users in North America would increase substantially.[11]

(4) It is the opinion of the committee that the military should be able to develop doctrine, establish procedures, and train troops to operate in an L_1 jamming environment in less than three years.

---

[11] The analysis by Michael Dyment, Booz•Allen & Hamilton, 1 May 1995, is shown in Appendix E.

(5)   The technology for developing direct Y-code receivers is currently available and the development and initial deployment of these receivers could be accomplished in a short period of time if adequately funded.

(6)   The FAA's WAAS, the Coast Guard's differential system, and GLONASS are expected to be fully operational in the next 1 to 3 years. The Coast Guard's DGPS network and the WAAS will provide accuracies greater than that available from GPS with SA turned to zero and GLONASS provides accuracies that are comparable to GPS without SA. At the same time, other local DGPS capabilities are likely to continue to proliferate.

Selective Availability should be turned to zero immediately and deactivated after three years. In the interim, the prerogative to reintroduce SA at its current level should be retained by the National Command Authority.

Anti-Spoofing

The purpose of A-S is to protect military receivers from an adversary transmitting a spoofed P-code signal and to deny the precision to an adversary through encryption.[12] When A-S is turned on, the P-code modulation on both the L_1 and L_2 carriers is replaced with a classified known as the Y-code that has the same chipping rate and correlation properties as the P-code. (C/A-code is not affected by Y-code transmission.) Except for special arrangements to turn off A-S for specific requirements, it has remained on continuously since January 31, 1994.

Impact of A-S on Military Users

PPS receivers are able to track the Y-code through the use of a security module that employs National Security Agency cryptographic techniques, and requires the manual distribution of encryption keys.

There are compelling reasons to retain the A-S feature. If the recommendation to remove SA is implemented and potential adversaries have access to the resulting more accurate C/A-code on the L_1 frequency, the reasons to retain A-S become still more compelling. In addition to its anti-spoofing feature, A-S forces adversaries to use the C/A-code on the L_1 frequency, which can be denied by jamming techniques (without impacting L_2). The NRC committee believes that denying L_1 to an enemy through jamming, while employing only L_2 for its own forces, should be the basis of a new military doctrine for the use of GPS. However, this doctrine will require U.S. military receivers to acquire the Y-code rapidly without the C/A-code. Military receivers also should be able to provide accurate

---

[12] The process of sending incorrect information to an adversary's radio equipment (in this case a GPS receiver) without their knowledge, using mimicked signals, is known as spoofing.

ionospheric corrections in the absence of L_1. Modifications to military receivers to accomplish this are discussed later in this report.

The current manual distribution of decryption keys is laborious and time consuming. The DOD is currently developing the means to distribute the keys electronically. Such a capability would greatly enhance the use of the encrypted L_2 Y-code. The committee also believes that technology is available to upgrade the current P-code encryption method and suggests that the Air Force should explore the necessity of utilizing this technology. Modifications to the Block IIR satellites and the Block IIF request for proposal may be required if upgraded encryption methods are necessary. Changes to military receivers also will be required.

Impact of A-S on Civil Users

SPS receivers cannot directly track Y-code, which significantly limits civil access to measurements of the L_2 signal for correction of ionospheric errors. Several Y-codeless approaches have been developed to overcome this problem.[13] These techniques, however, have a lower signal-to-noise ratio than dual-frequency tracking. This creates difficulties in situations where the receiver is moving, is subject to multipath signals, or is operating in areas where signal attenuation exists, such as in an urban area or under foliage. Despite these limitations, less approaches are still being used for many surveying and scientific applications. SPS users also would benefit from access to an unencrypted L_2 signal, because its bandwidth is approximately ten times as wide as the L_1 signal. The wider bandwidth would improve resistance to interference and reduce vulnerability to multipath.

Findings and Recommendations

A-S is critically important to the military because it forces potential adversaries to use the C/A-code on L_1, which can be jammed if necessary without inhibiting the U.S. military's use of the encrypted Y-code on L_2. Further, encryption provides resistance to spoofing of the military .

Although many civil users could benefit if A-S is turned off, as discussed in the previous chapter and above, their requirements can be met with other enhancements described in subsequent pages.

A-S should remain on and the electronic distribution of keys should be implemented at the earliest possible date. In addition, the Air Force should explore the necessity of upgrading the current encryption method. Required receiver enhancements should be incorporated in future planned upgrades.

---

[13] Some codeless approaches include (1) delay and multiply to recover the carrier and code phases, (2) squaring to recover the carrier phase, (3) cross-correlation of the L_1 and L_2 signals to measure the differential carrier phase and code pseudorange, and (4) P-code enhanced versions of these techniques.

SIGNAL STRUCTURE MODIFICATIONS TO REDUCE
ATMOSPHERIC DELAY ERROR

If SA is turned to zero, as recommended above, the next largest contributor to the civilian SPS error budget is the atmospheric error consisting mainly of ionospheric delay as discussed in Appendix C and as shown in Table 3-1. Since the military normally has access to two frequencies, military users can correct the ionospheric error,[14] but civilian users cannot.[15] In order to compensate for the ionospheric error, the civilian community has been able to develop innovative techniques for recovering components of the encrypted Y-code signal. The chief limitation on the use of these somewhat expensive receivers is that to function effectively, the signal-to-noise ratio required for the L_2 signal must be considerably higher than that required by a military PPS receiver. While this is achievable in stationary situations, there are many circumstances in which these conditions do not apply. For example, when the receiver is in a moving vehicle and/or there is ionospheric scintillation present (which corrupts the phase of the received signals), the receiver can lose lock. Several minutes may be required to recover the tracking ambiguity cycle needed for precise positioning.[16] The same is true when the receiver must view some of the satellites through foliage or in the presence of multipath signals.

Unfortunately, an ideal signal reception environment is the exception rather than the rule. As the number of more demanding real-time civil applications increases, users are seeking ways to improve GPS performance. With the current GPS signal structure, civilian designers and users must confront impediments such as non-trivial levels of electrical interference and strong and rapidly changing multipath reflections from buildings and nearby vehicles. Civilian access to an additional frequency would enable improved accuracy through ionospheric corrections, multipath rejection, and single-frequency operation when interference jams one of the two civilian frequencies.

---

[14] As mentioned in Appendix C, the tropospheric portion of atmospheric delay cannot be eliminated through the use of two frequencies.

[15] Because the ionosphere is a dispersive medium, the ionospheric delay is frequency dependent. The existence of two frequencies allows the time of arrival of each to be compared by a receiver, calibrating the error caused by signal delay through the Earth's ionosphere. PPS users have access to both L_1 and L_2, whereas SPS users have access only to L_1.

[16] Ionospheric scintillation of the GPS signals occurs when two or more paths are taken between the satellite and the receiver. This is caused by fluctuations in the free electron content and therefore, the refractivity of the ionosphere. When these paths carry signals of about the same amplitude, they cancel as the differential delay of the paths vary by integer plus one-half wavelengths, or they add as the differential delay of the paths vary by integer wavelengths. This scintillation is analogous to optical delays in the neutral atmosphere, which cause stars to twinkle in the visible spectrum.

Guidelines and Technical Considerations

In studying possible options for the addition of another civilian frequency, a set of guidelines and technical considerations was developed as follows:

(1)   The signal must not interfere with the military's jamming techniques for denial of GPS signals.

Any signal enhancement should preserve and maximize the ability of the military to deny the GPS signal to adversaries through local jamming of any unencrypted s without adversely impacting the L_2 Y-code signal. The use of encryption on the Y-code effectively denies its use to unauthorized parties.

(2)   The signal must be backward compatible.

A significant investment has been made in receiver purchases and existing receiver performance must not be degraded; although existing receivers may not be able to take advantage of the new signal.

(3)   The frequency allocation for the signal must be considered.

The signal should be assigned a frequency in the L-band spectrum that has a reasonable chance of receiving an official allocation from the Federal Communications Commission and, in some cases, the International Telecommunications Union as well. By using an L-band frequency, the cost of receiver modifications should not increase substantially.[17] Unfortunately however, because many of the proposed mobile satellite communication services (Iridium, Globalstar, and others) plan to use L-band frequencies, L-band frequency allocation is difficult to attain. In light of this potential problem, a preliminary assessment was undertaken to identify possible L-band frequencies that could be used for transmission of an additional GPS signal.[18] Based on this preliminary assessment, it appears that several sub-bands have promise for the proposed signal, and several frequencies were selected as potential candidates. Although these frequencies are included in Table 3-5, in-depth investigation and coordination will be required before a specific frequency band, wide or narrow, can be selected.

---

[17] The addition of an L_4 signal would not affect the operation of existing receivers, but manufacturers would have to modify future receivers (add another channel, and change the correlator and processor) to take advantage of a new L-band signal. If a frequency much greater than L-band is used, additional antennas would have to be added to the receivers, and the satellite transmitted power would have to increase.

[18] A preliminary analysis of the L-band spectrum allocation that was conducted by Mr. Melvin Barmat, Jansky/Barmat Telecommunications Inc., Washington D.C., January 1994, is shown in Appendix I.

(4)   *The signal should optimally be spaced for ionospheric correction and wide lane ambiguity resolution.*

The NRC committee determined that ideally, the new GPS signal should be on an L-band frequency sufficiently offset from L_1 to permit user correction of ionospheric delay, which would improve user accuracy yet be close enough to L_1 to allow fast, wide-lane cycle ambiguity resolution, also termed wide-laning.[19] For adequate ionospheric correction, the separation between L1 and a new frequency should be at least 200 MHz.[20] For optimal wide-lane ambiguity resolution, the frequency difference between L_1 should be no greater than 350 MHz.

(5)   *The signal should occupy a wide frequency band.*

The signal should occupy a wide frequency band, that is, around 10 MHz, to reduce the effects of multipath and improve resistance to unintentional RF interference. A wide-band signal has two main advantages over a narrow-band signal.[21] First, use of a wide-band signal allows about a 10-dB improvement in interference rejection over a narrow-band signal. This is significant for both stand-alone and differential users needing improved availability in the presence of wide-band or continuous wave interference. The second advantage is that upon signal reacquisition, a wide-band signal can recover submeter pseudorange accuracy faster than a narrow-band signal in both low- and high-multipath environments. For example, as discussed in Appendix G, in a high multipath environment, such as around buildings, a narrow-band signal will have an error larger than a wide-band signal after signal reacquisition. Many important real-time vehicular applications, such as aircraft precision approach and land vehicle guidance, would benefit from the faster accuracy recovery obtained with a wide-band, faster chipping-rate signal.

New Signal Structure Options

Ten signal structure enhancements to the current GPS signal structure were considered and are described in Appendix H. Each option involved possible changes to L_1 or L_2, as well as possible signal transmissions on a new frequency. Using the previously

---

[19] Wide-lane ambiguity resolution (wide-laning) is a processing technique developed by civilian DGPS users to process carrier phase data. With wide-laning, the two carrier frequencies are mixed to provide a difference frequency of about 4.5 times longer wavelength, improving the speed and reliability of cycle ambiguity resolution. The wide-laning technique is available to cross-correlation types of receivers today, but at a serious loss in effective carrier-to-noise ratio as compared with a dual-frequency code-tracking receiver.

[20] Letter from J. A. Klobuchar, U.S. Air Force Geophysics Laboratory, 22 December 1994.

[21] A wide-band signal is generally defined to be around 20 MHz wide; a narrow-band signal around 2 MHz wide.

discussed guidelines, the NRC committee determined that 2 of the 10 options should be seriously considered. These two options are discussed below in order of preference.

Option 1: Wide-Band L_4 Signal

The optimal scenario for an enhanced civilian GPS signal would entail the provision of a new wide-band frequency, termed L_4, that would be broadcast unencrypted to allow for universal access. The wide bandwidth sufficiently offset from the current L_1 signal would allow for ionospheric delay correction, wide-lane ambiguity resolution, improved interference rejection, and faster accuracy recovery in multipath environments.

The pseudorandom noise chosen for the L_4 wide-band signal should have a bandwidth similar to the present P-code, but with a sequence length chosen for rapid acquisition by low-cost civilian receivers.[22] Although not needed for acquisition purposes, the signal could have C/A-code in phase quadrature, which would allow manufacturers to get the most benefit from the new signal without significant changes to their investment in application-specific integrated circuit (ASIC) correlators.[23]

Based on the previously mentioned frequency allocation analysis, it appears that several options may exist for a wide-band L_4 signal. The first option would be to place the center of the wide-band L_4 signal at 1258.29 MHz. If the Russian Federation follows through on plans to move GLONASS L_2 transmissions to the lower portion of their frequency allocation (1242.9-1251.7 MHz by 1998 and 1242.9-1248.6 by 2005), even a wide-band signal placed at 1258.29 MHz would cause little frequency overlap. Therefore, the possibility of interference with GLONASS would be low. The second option would be to place the wide-band L_4 signal at 1841.40 MHz. Again, the feasibility of receiving a frequency allocation in this area of the spectrum would require further investigation.

Option 2: Narrow-Band L_4 Signal

If a wide-band frequency allocation proves impossible to obtain for L_4, a narrow-band signal should be considered as the second best option. Several potential frequencies have been identified that have sufficient spacing from L_1 to allow for the correction of ionospheric delay. These include 1237.83 MHz (which is the upper null of the existing L_2 frequency); 1258.29 MHz; and 1841.40 MHz. A narrow-band signal placed at any of these frequencies would carry a C/A-type code.

---

[22] The code sequence length for the current P-code is 1 week.

[23] A dual-frequency L_1/L_4 receiver would still need an additional RF/IF (intermediate frequency) section and synthesizers. For current dual-frequency receiver manufacturers, hardware changes would not be difficult.

Additional Considerations

Regardless of which frequency, bandwidth, and type is chosen for the new L_4 signal, its relative utility to a number of different user communities also will be affected by the type of data superimposed on the signal. For example, the inclusion of integrity information in a data message would be useful to aviation, maritime, and land transportation users concerned with safety. A navigation message also would be useful because it would allow the L_4 signal to be used for navigation without access to an additional frequency (that is, L_1 or L_2). Users employing codeless techniques, who are interested in improved correlation between the L_4 signal and the L_1 signal, would benefit from having the same data transmitted on each signal. However, if a navigation message were broadcast unencrypted, potential adversaries of the United States also could take advantage of an L_4 signal in a theater of war, unless L_4 is jammed along with other radionavigation signals. Thus, an L_4 signal with no data would probably be most acceptable from a military perspective.

The rate at which data are broadcast on the L_4 signal also is important. A high data rate would increase the amount of information that could be sent to a user and would allow the information to be sent very quickly. High data rates, however, generally make a signal more susceptible to jamming. Conversely, a signal with a low data rate is more jam resistant, but also is limited in its ability to get information to a user in a timely manner. Data rate also may have an impact on the power level required for a new L_4 signal, which is an important consideration because of its effect on required satellite power.

Because of these many considerations, the committee believes that it is premature to suggest a specific data message or broadcast rate for the L_4 signal, but believes that it should be designed with the flexibility to add the data considered most critical to the GPS user community when the first L_4-capable satellite is launched.

Improvements Anticipated from Adding L_4

Increased Accuracy

The new L_4 signal, which would be available to civilian users, would reduce the typical ionospheric error of 7.0 meters to 0.01 meters (1σ), regardless of the option selected, as shown in Table 3-5. This would result in a stand-alone accuracy as low as 21.2 meters (2 drms) compared with approximately 30 meters (2 drms) with L_1 alone. With the addition of the L_4 signal, several DPGS accuracy requirements could be met with the stand-alone GPS accuracy, including those for surface surveillance and autonomous vehicle location and interrogation. The addition of an L_4 signal also assists short- and long-baseline differential users (e.g., Category III approach and landing, mapping, surveying, precision farming, and Earth science applications) by calibrating the spatially uncorrelated components of the ionosphere seen across the baseline, and by speeding up ambiguity resolution to get accuracies of a decimeter or better. Even in the presence of SA, dual-frequency civil receivers that operate in a codeless mode would benefit from an additional, unencrypted, signal.

Additional Benefits of L₄

The existence of an unencrypted L_4 signal greatly reduces a civilian receiver's probability of RF interference by providing a second frequency, which can be used in the event that L_1 is subject to interference. The wide-band L_4 signal also would aid in commercially important emerging markets where reception is less than ideal, since GPS must operate in applications subject to strong and intermittent multipath and signal blockage. The success or failure of GPS in those applications depends upon quick recovery of accurate pseudorange measurements once the signal is restored.

From the military perspective, the addition of the L_4 signal retains A-S on both L_1 and L_2 and is quite flexible with respect to selective denial of civilian service. Of all the frequencies mentioned above, 1237.83 MHz would be the most difficult to jam because it is the closest to L_2. However, based on an analysis described in Appendix J, this frequency could be selectively jammed without affecting the use of the Y-code on L_2. In order to selectively deny civilian service, broadband jamming of L_1 and L_4 could be used. Note that even if no navigation message is broadcast on the L_4 signal, it should be jammed because the last ephemeris information could be used in combination with L_4 ranging data to locate a target. It also should be noted that broadband jamming of both L_1 and L_4 would eliminate the capability for dual-frequency ionospheric corrections. This would reduce PPS accuracy and force the U.S. military to rely on other methods of obtaining ionospheric corrections. As discussed later in this chapter, ionospheric correction models broadcast on the navigation message remove only about 50 percent of the ionospheric error. However, by using receivers with the capability to store the last known ionospheric correction and updating that information with a process called Differential Ranging Versus Integrated Doppler (DRVID), ionospheric corrections can be improved further over the 50 percent correction obtained in the L_2 broadcast models.

Reduction of Receiver Noise and Multipath Errors

As shown in Table 3-5, when using a typical SPS receiver, the receiver noise and multipath actually increase when another frequency is added because of the noise and multipath from the additional frequency. As a result, the beneficial effects of adding another frequency to reduce the ionospheric error are diminished. If more advanced receivers are used, reductions in the receiver noise and multipath errors can be achieved, and the HDOP can be reduced to around 1.5.[24] The error reductions achieved by using a more advanced receiver results in stand-alone SPS performance ranging from 11.3 meters to 13.1 meters (2

---

[24] The characteristics of a more advanced, dual-frequency SPS and PPS receivers (as compared to the typical receiver described previously) include: (1) use of more satellite signals in the solution (typically six to eight satellites), (2) lower noise amplifier, (3) better tropospheric model, (4) on-board multipath processing capability and low-multipath antenna, and (5) lower C/A-code measurement noise due to narrow correlator spacing. For an all-in-view receiver and a elevation mask angle of 5 degrees, an HDOP of 1.5 is predicted 95 percent of the time. Source: Analysis completed by Mr. Tom Hsiao of the MITRE Corporation, 15 February 1995.

drms), depending on the L_4 signal bandwidth and frequency, as shown in Table 3-6. These accuracies would satisfy the accuracy requirements for aviation traffic alert/collision avoidance systems (TCAS). The PPS performance would improve to 11.1 meters (2 drms) or 4.6 meters (CEP), as shown in Table 3-7.

With accuracy levels of 11.3 to 13.1 meters (2 drms), GPS availability also is enhanced, and RAIM is improved as well. For example, for a stand-alone horizontal accuracy of 100 meters, the availability of four satellites would increase from the previous value of 99.94 percent to approximately 99.96 percent. RAIM availability, which is dependent on the presence of six useable satellite signals, is shown in Table 3-8.

Although not shown in Tables 3-6 or 3-7, even further improvements to the receiver noise and multipath errors can be made through use of the most advanced receivers that have improved receiver signal processing, are integrated with auxiliary sensors, and have multi-element antenna arrays.

Table 3-5 Elimination of Ionospheric Error by the Addition of Another Frequency.

Error Source	SPS With II/IIA Satellites	Typical Range Error Magnitude (meters, 1σ)				
		SPS Improved (no SA, additional narrow L-band signal)			SPS Improved (no SA, additional wide L-band signal)	
		1237.83	1258.29	1841.40	1258.29	1841.40
		Narrow-band, C/A-type code	Narrow-band, C/A-type code	Narrow-band, C/A-type code	Wide-band, P-type code	Wide-band, P-type code
Selective Availability	24.0	0.0	0.0	0.0	0.0	0.0
Atmospheric Error Ionospheric Tropospheric	7.0 0.7	0.01 0.7	0.01 0.7	0.01 0.7	0.01 0.7	0.01 0.7
Clock and Ephemeris Error	3.6	3.6	3.6	3.6	3.6	3.6
Receiver Noise	1.5	4.6	4.9	6.9	2.7	5.6
Multipath	1.2	3.7	3.9	5.6	2.7	4.8
Total User Equivalent Range Error (UERE)	25.3	6.9	7.3	9.6	5.3	8.2
Typical Horizontal DOP (HDOP)	2.0	2.0	2.0	2.0	2.0	2.0
Total Stand-Alone Horizontal Accuracy (2 drms)	101.2	27.8	29.0	38.5	21.2	32.9

Table 3-6 Effect of Reduced Ionospheric Error by the Addition of Another Frequency and Additional Improvements Obtained with Using a More Advanced SPS Receiver[a]

Error Source	SPS With II/IIA Satellites	Typical Range Error Magnitude (meters, 1σ)				
		SPS Improved (no SA, additional narrow L-band signal)			SPS Improved (no SA, additional wide L-band signal)	
		1237.83	1258.29	1841.40	1258.29	1841.40
		Narrow-band, C/A-type code	Narrow-band, C/A-type code	Narrow-band, C/A-type code	Wide-band, P-type code	Wide-band, P-type code
Selective Availability	24.0	0.0	0.0	0.0	0.0	0.0
Atmospheric Error Ionospheric[b] Tropospheric[c]	7.0 0.7	0.01 0.2	0.01 0.2	0.01 0.2	0.01 0.2	0.01 0.2
Clock and Ephemeris Error	3.6	3.6	3.6	3.6	3.6	3.6
Receiver Noise[d]	1.5	0.6	0.7	0.9	0.5	0.8
Multipath[e]	1.2	1.5	1.6	2.3	1.0	1.9
Total User Equivalent Range Error (UERE)	25.3	3.9	4.0	4.3	3.8	4.2
Typical Horizontal DOP (HDOP)[f]	1.5	1.5	1.5	1.5	1.5	1.5
Total Stand-Alone Horizontal Accuracy (2 drms)	76.0	11.9	12.0	13.1	11.3	12.5

a.   The characteristics of a more advanced, dual-frequency SPS receiver (as compared to the typical receiver described previously) include: (1) use of more satellite signals in the solution (typically six to eight satellites), (2) lower noise amplifier, (3) better tropospheric model, (4) on-board multipath processing capability and low multipath antenna, and (5) lower C/A-code measurement noise due to narrow correlator spacing.

b.   With the addition of an unencrypted, coded signal, the SPS ionospheric error is removed by a linear combination of the L_1 and L_4 observables. This correction leaves residual ionospheric error of 1 centimeter or less.

c.   For improved receivers, software models correct for all but around 0.2 meters (1σ) of the tropospheric error.

d.   For an improved SPS receiver, the receiver noise for independent 1-second measurements can be as low as 0.2 m for the narrow-band signal, and 0.1 meter for the wide-band signal. These are the single-frequency errors and must be increased to account for the linear combination used to calibrate ionospheric errors. For example, the narrow-band error must be multiplied by a factor of 3.1 when 1237.83 MHz and 1575.42 MHz (L_1) frequencies are used.

e.   For an SPS receiver with a low-multipath antenna and on-board multipath reduction processing, the multipath can be as low as 0.5 meters (1σ) for the narrow-band signal, and 0.2 meters (1σ) for the wide-band signal. These errors are very dependent on the number of reflective objects near the antenna. These are the single-frequency errors and must be increased to account for the linear combination used to calibrate ionospheric errors. For example, the narrow-band error must be multiplied by a factor of 3.1 when 1237.83 MHz and 1575.42 MHz (L_1) frequencies are used.

f.   For an all-in-view receiver and a elevation mask angle of 5 degrees, an HDOP of 1.5 or less was predicted 95 percent of the time. Source: Analysis completed by Mr. Tom Hsiao, the MITRE Corporation, 15 February 1995.

Table 3-7 Effect of Using a More Advanced PPS Receiver on Stand-Alone Accuracy[a]

Error Source	Typical Range Error Magnitude (meters, 1σ)	
	PPS with Typical Receiver	PPS with Advanced Receiver
Selective Availability	0.0	0.0
Atmospheric Error Ionospheric[b] Tropospheric[c]	0.01 0.7	0.01 0.2
Clock and Ephemeris Error	3.6	3.6
Receiver Noise[d]	0.6	0.3
Multipath[e]	1.8	0.6
Total User Equivalent Range Error (UERE)	4.1	3.7
Typical Horizontal DOP (HDOP)[f]	2.0	1.5
Total Stand-Alone Horizontal Accuracy, 2 drms	16.4	11.1

a.   The characteristics of a more advanced, dual-frequency PPS receiver (as compared to the typical receiver described previously) include: (1) use of more satellite signals in the solution (typically six to eight satellites), (2) lower noise amplifier, (3) better tropospheric model, and (4) on-board multipath processing capability and low-multipath antenna.

b. For a PPS receiver, the ionospheric error is removed by a linear combination of the L_1 and L_2 observables. This correction leaves residual ionospheric error of 1 centimeter or less.

c. For improved PPS receivers, software models correct for all but around 0.2 meters (1σ) of the tropospheric error.

d. For an improved PPS receiver, the receiver noise for independent 1-second measurements can be as low as 0.1 meters (1σ). These are the single-frequency errors and must be increased to account for the linear combination used to calibrate ionospheric errors. The single-frequency error of 0.1 meters must be multiplied by a factor of 3 when the standard $L_2 = 1227.6$ MHz and $L_2 = 1575.42$ MHz frequencies are used.

e. For an improved PPS receiver with a low-multipath antenna and on-board multipath reduction processing, the multipath can be as low as 0.2 meters (1σ). These errors are very dependent on the amount of reflective objects near the antenna. These are single-frequency errors and must be increased to account for the linear combination used to calibrate ionospheric errors. For example, the single-frequency error of 0.2 m must be multiplied by a factor of 3 when the standard $L_2 = 1227.6$ MHz and $L_1 = 1575.42$ MHz frequencies are used.

f. For an all-in-view receiver and a elevation mask angle of 5 degrees, an HDOP of 1.5 or less was predicted 95 percent of the time. Source: Analysis completed by Mr. Tom Hsiao, the MITRE Corporation, 15 February 1995.

Table 3-8 Effect of SA Removal and Dual-Frequency Capability on RAIM Availability for Aviation Applications[a]

Aviation Application		Availability With SA Set to Zero		Availability With SA Turned to Zero and L_4 Added	
Phase of Flight	Protection Limit	21 Satellites[b]	24 Satellites[c]	21 Satellites	24 Satellites
En Route	2.0 nautical miles	96.34%	99.98%	96.80%	100.00%
Terminal Area	1.0 nautical miles	94.39%	99.95%	95.19%	99.98%
Non-precision Approach	0.3 nautical miles	91.10%	100.00%[d]	93.12%	100.00%[d]

a. This analysis has been made for a single frequency C/A-code receiver aided by a barometric altimeter (required for aviation supplemental navigation use of GPS) with a visibility mask angle of 5 degrees.

b. The probability of having 21 satellites operating is assumed to be 98 percent.

c. The probability of having 24 satellites operating is assumed to be only 70 percent. However, the values in this table reflect the fact that if 24 satellites are fully operational, an incremental improvement in availability exists.

d. Although these values would intuitively be lower than the 1 nautical mile terminal area protection limit value, availability improves for the 0.3 nautical mile non-precision protection limit because the barometric altimeter inputs provide extra information in this phase of flight.

Findings and Recommendations

The NRC committee determined that the addition of a new, L-band signal, L_4, offers civilian users much improved precision in many reception environments as well as preserving selective denial options for the military. The NRC committee anticipates that domestic suppliers of commercial GPS receivers, who also are the suppliers of dual-frequency military receivers, would enjoy some advantage over foreign competitors in providing dual-frequency civilian receivers.

The NRC committee believes that the L_4 signal could be added to several Block IIR spacecraft using the existing volume and power on the Block IIR spacecraft. If it is assumed that the L_4 signal transmits at a radiated power similar to the L_1 or L_2 signals, then approximately 180 watts of DC power is required.[25] The exact amount of power however, will depend on the specific frequency selected for L_4. Since the current Block IIR L-band (L_1, L_2, and L_3) navigation payloads and harnesses weigh around 160 kilograms (353 lbs), the L_4 signal generation system is expected to weigh approximately one-fourth to one-fifth that amount.[26] Based on information provided to the NRC committee through various presentations, it is believed that the sufficient power for an additional frequency can be made available on the Block IIR spacecraft by utilizing the currently unused Reserve Auxiliary Payload power margin, and by re-definition and re-allocation of other existing margins.

In order to add a new signal, several Block IIR hardware modifications are required, including the addition of a frequency synthesizer, modulator/intermediate power amplifier, a high-power amplifier, and a payload processor.[27] The NRC committee believes that adequate space for this additional hardware currently exists on the Block IIR spacecraft. Based on cost information for the current Block IIR L-band navigation package, the committee believes that the addition of another, unencrypted L-band signal would cost approximately $1.3 million per Block IIR satellite.[28]

Immediate steps should be taken to obtain authorization to use an L-band frequency for an additional GPS signal, and the new signal should be added to GPS Block IIR satellites at the earliest opportunity.

---

[25] Information provided by Martin Marietta Astro Space Division of Lockheed-Martin, 6 February 1995.

[26] Information provided by Martin Marietta Astro Space Division of Lockheed-Martin, 12 April 1995.

[27] Information provided by Martin Marietta Astro Space Division of Lockheed-Martin, 6 February 1995 and by ITT Corporation, 13 March 1995.

[28] It is estimated that the non-recurring design and development costs for each of the existing Block IIR L-band signals are $11 million, and the unit price for each existing L-band signal is around $500,000 per satellite. It is estimated that the cost for each L_4 signal payload processor would be $100,000, and the non-recurring costs for deliverable test equipment would be $3 million. Information provided by ITT Corporation, 13 March 1995.

PERFORMANCE IMPROVEMENTS TO THE GPS OPERATIONAL CONTROL SEGMENT AND SATELLITE CONSTELLATION

Current Status of the Operational Control Segment and Planned Upgrades

The current operational control segment (OCS) consists of a master control station (MCS) plus four additional monitor sites that collect GPS Y-code measurements from a maximum of 11 satellites each.[29] All but one of these sites are capable of sending uploads to the GPS satellites.

There are plans to award an OCS consolidated contract in July 1995 to provide maintenance until the year 2000, to make improvements to the existing software architecture and user interfaces, and to support the deployment and operation of the Block IIR satellites. There is an option in the contract to replace the operational control software, but the winning contractor can choose to upgrade existing software rather than replace it.

In February 1996, the Air Force plans to award another OCS contract, which will be effective beginning in the year 2000. This contractor will assume responsibility for the operational control segment, the Block IIR constellation, and the development and deployment of the Block IIF satellites. However, neither of these contracts address critical upgrades that would enhance the operation of the OCS and thus enhance the performance of GPS.

The NRC committee recommends changes below that will enhance the overall GPS operation and improve performance. Most of these changes focus on the OCS and can be implemented immediately. Some improvements, however, focus on the operation of the Block IIR constellation and cannot be introduced until several Block IIR satellites are in orbit.

Recommended Upgrades to the Operational Control Segment

In addition to other operational functions, such as satellite health monitoring and routine maintenance, the GPS control segment is responsible for determining the ephemeris[30] and clock parameters and uploading them to the satellites. A partitioned Kalman Filter[31] at the master control station estimates the orbits and clock errors for each

---

[29] Information provided by Air Force Space Command, 1 December 1994.

[30] Ephemeris is defined as a satellite's position as a function of time.

[31] A Kalman Filter incorporates both observations and mathematical models of the system dynamics to produce an estimate of the current state of a system. By using knowledge of how the system state can change over time, the Kalman Filter allows the contributions of individual measurement errors to be averaged. In the MCS filter, the system state includes satellite orbital parameters, clock parameters, and numerous other elements.

of the GPS satellites as well as the clock errors for the monitor site receivers. Updated orbit and clock corrections are uploaded to each satellite once a day.

With the current GPS constellation, the clock and ephemeris errors contribute approximately 3.4 and 1.4 meters (1σ), respectively, to the SPS and PPS error budget, for a combined error of 3.6 meters (1σ)[32], as shown in Table 3-1.[33] Once SA, the atmospheric, receiver noise, and multipath errors have been eliminated or reduced, ephemeris and clock errors become the largest contributors to the UERE. As shown below, several methods can be used to reduce combined clock and ephemeris errors to increase accuracy and improve overall performance.

Accuracy Improvements

Planned Experiments Involving Correction Updates and Additional Monitoring Stations. An innovative, near-term strategy for improving PPS accuracy and integrity has been investigated by the Air Force, and an experiment to test the strategy is expected to begin in the spring of 1995. The experiment involves uploading pseudorange corrections for all satellites with each scheduled, individual satellite upload.[34] These corrections would be made available to PPS users in the navigation message. A PPS receiver can decode the messages from all satellites it is tracking and apply the most recent correction set. The Air Force expects that this will improve the combined error contribution of clock and ephemeris for PPS users by half, to approximately 2 meters (1σ). If SA is turned to zero as previously recommended, SPS users will not receive the same benefit from this experiment as PPS users unless current security classification policies are changed to allow the most recent clock and ephemeris parameters to be broadcast from each satellite unencrypted.[35]

In conjunction with the above experiment, the Air Force is investigating another enhancement that could provide further reduction in the combined PPS clock and ephemeris error. This enhancement involves the integration of data from five Defense Mapping Agency (DMA) GPS monitoring sites with the existing Air Force operational control segment in a simulated Kalman Filter. By including additional data from the DMA sites, which are located at higher latitudes than the Air Force sites, an additional 15 percent improvement in combined clock and ephemeris accuracy can be anticipated, based on tests previously

---

[32] The error of 3.6 meters (1σ) was obtained by taking the square root of the sum of the squares of 3.4 and 1.4 meters (1σ).

[33] J. F. Zumberge and W. I. Bertiger, "Ephemeris and Clock Navigation Message Accuracy in the Global Positioning System," Volume I, Chapter 16. Edited by B. W. Parkinson, J. J. Spilker, P. Axelrad, and P. Enge. To be published by AIAA, in press, 1995.

[34] Satellites are normally uploaded once per day.

[35] Currently the most recent clock and ephemeris updates are broadcast in an encrypted portion of the navigation message. Clock and ephemeris parameters less than 48 hours old are classified.

conducted by the DMA.[36] It should be emphasized that this experiment will be conducted as a software simulation only, so PPS users will not actually observe the additional 15 percent simulated improvement.

Recommended Implementation of More Frequent Updates and Additional Monitoring Stations. Full operational implementation of the first experiment, which involves uploading of clock and ephemeris corrections for all satellites with each scheduled, individual satellite upload, should not be difficult to accomplish and would appear to reduce the combined clock and ephemeris error to half of its current value.

Operational implementation of the second planned experiment, which involves the incorporation of DMA monitor site data, is more difficult to achieve. While well-distributed geometrically, DMA GPS monitor stations do not have secure communications data links to the master control station. Existing Air Force sites, which are used for other purposes, have secure data links to Air Force Space Command (co-located with the GPS master control station), but are not well distributed in latitude for GPS monitoring and do not have GPS receivers. Additional GPS monitoring sites are expected to improve stand-alone GPS accuracy. More importantly, a well-distributed set of monitor sites would allow continuous tracking of each satellite, enabling the prompt detection of satellite failures. An estimated cost of $9 million for using DMA data in real-time and an estimated cost of co-locating Air Force monitor stations at DMA sites was provided to the committee.[37]

> ***The DOD's more frequent satellite navigation correction update strategy should be fully implemented as soon as possible following the successful test demonstration of its effectiveness. In addition, the current security classification policy should be examined to determine the feasibility of relaxing the 48-hour embargo on the clock and ephemeris parameters to civilian users.***

> ***Additional GPS monitoring stations should be added to the existing operational control segment. Comparison studies between cost and location should be completed to determine if Defense Mapping Agency or Air Force sites should be used.***

Recommended Use of a Non-Partitioned Kalman Filter with Improved Dynamic Models. The original computer hardware used for the OCS was not capable of processing all satellites in a single Kalman Filter. The existing software was written with this limitation as well. The hardware has since been upgraded, leaving only the software to restrict full processing of all satellite clock and ephemeris data simultaneously. Unfortunately, there currently are no definite plans to upgrade the Kalman Filter software, including the dynamic

---

[36] Stephen Malys, DMA, Viewgraphs from presentation at the PAWG 1993, Colorado Springs, Colorado.

[37] Information provided by the Aerospace Corporation, 21 February 1995.

model. It is possible that the winning contractor of the 1995 contract may choose to eliminate the partitions, but there is not a specified requirement to do so.

Based on recent improvements to the DMA's Kalman Filter, which originally had a configuration similar to the GPS Kalman Filter, use of an updated, non-partitioned GPS Kalman Filter is expected to reduce the combined clock and ephemeris error by 15 percent.[38] Furthermore, an additional 5 percent improvement can be achieved by using improved dynamic models in the Kalman Filter, which would allow better predictions of satellite behavior 1 day ahead.[39] An estimated cost of $7.5 for upgrading the Kalman Filter and improving its dynamic models was provided to the committee.[40]

> ***The operational control segment Kalman Filter should be improved to solve for all GPS satellites' clock and ephemeris errors simultaneously through the elimination of partitioning and the inclusion of more accurate dynamic models. These changes should be implemented in the 1995 OCS upgrade request for proposal.***

The combined clock and ephemeris improvement obtained with each of the above upgrades is shown in Table 3-9. If all three of the recommendations above are implemented, the combined clock and ephemeris error is expected to be approximately 1.2 meters (1σ). As shown in Table 3-10 and Figure 3-5, if: (1) SA is turned to zero; (2) an additional GPS L-band signal is added; (3) more advanced receivers are utilized; and (4) each of the clock and ephemeris accuracy improvements are implemented, then a stand-alone GPS SPS accuracy of 5.4 meters (2 drms) with a narrow, L-band signal should be obtainable, and a stand-alone GPS SPS accuracy of 4.9 meters (2 drms) with a wide-band signal should be obtainable.[41] In addition, as shown in Table 3-11, a PPS accuracy of 4.2 meters (2 drms) (1.8 meters CEP) also would be obtainable.

With stand-alone accuracies at this level, many civilian and military accuracy requirements, such as the following will be met:

- Aviation — Category I approach and landing.

- Maritime — Recreational boating, vessel-tracking services, and harbor/harbor approach requirements.

---

[38] Stephen Malys, DMA, Viewgraphs from presentation at the PAWG 1993 meeting, Colorado Springs, Colorado.

[39] Stephen Malys, DMA, Viewgraphs from presentation at the PAWG 1993 meeting, Colorado Springs, Colorado.

[40] Information provided by the Aerospace Corporation, 21 February 1995.

[41] Civil users would have access to this level of accuracy only if the 48-hour embargo on clock and ephemeris parameters is lifted.

- ITS — Infrastructure management, highway navigation and guidance, mayday incident and alert, automated bus/railstop annunciation, collision avoidance (hazardous situation), and vehicle or cargo location (hazardous material transport).

- Earth Science — Oceanographic navigation and real-time positioning.

- Spacecraft — Real-time satellite orbit determination.

- Military — Precision-guided munitions.

Table 3-9 Reduction of Combined Clock and Ephemeris Errors

Enhancement	Anticipated Combined Clock and Ephemeris Error Improvement over Existing Combined Error of 3.6 meters (1σ)
Correction Updates (50% reduction)	1.8 meters
Additional Monitor Stations (additional 15% reduction)	1.5 meters
Non-partitioned Kalman Filter (additional 15% reduction)	1.3 meters
Improved Dynamic Model (additional 5% reduction)	1.2 meters

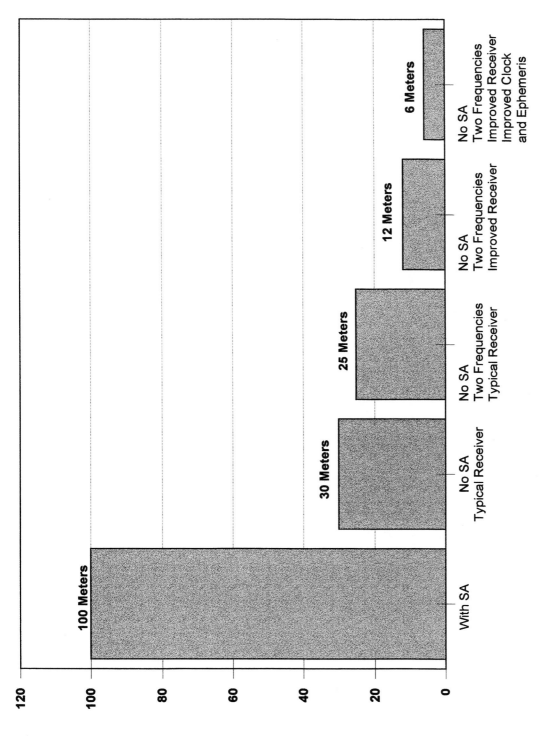

Figure 3-5 Approximate stand-alone horizontal SPS accuracy, 2 drms, resulting from recommended improvements and enhancements.

Table 3-10 Impact of Reduced Clock and Ephemeris Error on SPS Stand-Alone Accuracy

Error Source	SPS With II/IIA Satellites	Typical Range Error Magnitude (meters, 1σ)				
		SPS Improved (no SA, additional narrow L-band signal)			SPS Improved (no SA, additional wide L-band signal)	
		1237.83 Narrow-band, C/A-type code	1258.29 Narrow-band, C/A-type code	1841.40 Narrow-band, C/A-type code	1258.29 Wide-band, P-type code	1841.40 Wide-band, P-type code
Selective Availability	24.0	0.0	0.0	0.0	0.0	0.0
Atmospheric Error						
Ionospheric	7.0	0.01	0.01	0.01	0.01	0.01
Tropospheric	0.7	0.2	0.2	0.2	0.2	0.2
Clock and Ephemeris Error	3.6	1.2	1.2	1.2	1.2	1.2
Receiver Noise	1.5	0.6	0.7	0.9	0.5	0.8
Multipath	1.2	1.2	1.6	2.3	1.0	1.9
Total User Equivalent Range Error (UERE)	25.3	1.8	2.1	2.8	1.7	2.4
Typical Horizontal DOP (HDOP)	1.5	1.5	1.5	1.5	1.5	1.5
Total Stand-Alone Horizontal Accuracy (2 drms)	76.0	5.4	6.4	8.3	4.9	7.1

Table 3-11 Impact of Reduced Clock and Ephemeris Error on PPS Stand-Alone Accuracy

Error Source	Typical Range Error Magnitude (meters, 1σ)	
	PPS with II/IIA satellites	PPS Improved
Selective Availability	0.0	0.0
Atmospheric Error Ionospheric Tropospheric	0.01 0.2	0.01 0.2
Clock and Ephemeris Error	3.6	1.2
Receiver Noise	0.3	0.3
Multipath	0.6	0.6
Total User Equivalent Range Error (UERE)	3.7	1.4
Typical Horizontal DOP (HDOP)	1.5	1.5
Total Stand-Alone Horizontal Accuracy, 2 drms	11.1	4.2

As with the previous performance improvements, the increased positioning accuracy achieved by reducing clock and ephemeris errors also enhances availability. For example, for a stand-alone horizontal accuracy of 100 meters, the availability of four satellites would increase from the previous value of 99.96 percent to 99.97 percent. The improved RAIM availability is shown in Table 3-12.[42]

Table 3-12 Effect of SA Removal, Dual-Frequency Capability and Reduced Clock and Ephemeris Errors on RAIM Availability for Aviation Applications[a]

Aviation Application		Availability With SA Turned to Zero and L_4 Added		Availability With SA Turned to Zero, L_4 Added, and Reduced Clock and Ephemeris Error	
Phase of Flight	Protection Limit	21 Satellites[b]	24 Satellites[c]	21 Satellites	24 Satellites
En Route	2.0 nautical miles	96.80%	100.00%	97.08%	100.00%
Terminal Area	1.0 nautical miles	95.19%	99.98%	95.70%	100.00%
Non-Precision Approach	0.3 nautical miles	93.12%	100.00%[d]	94.36%	100.00%[d]

---

[42] Based on analysis conducted by the MITRE Corporation for the NRC committee, 7 February 1995. For more details, see footnote 1 earlier in this chapter.

a. This analysis has been made for a single-frequency C/A-code receiver aided by a barometric altimeter (required for aviation supplemental navigation use of GPS) with a visibility mask angle of 5 degrees.

b. The probability of having 21 satellites operating is assumed to be 98 percent.

c. The probability of having 24 satellites operating is assumed to be only 70 percent. However, the values in this table reflect the fact that if 24 satellites are fully operational, an incremental improvement in availability exists.

d. Although these values would intuitively be lower than the 1 nautical miles terminal area protection limit value, availability improves for the 0.3 nautical miles non-precision protection limit because the barometric altimeter inputs improve in this phase of flight.

Overall System Improvements

Improved Monitor Station Receivers. The receivers currently used at the monitor stations are outdated compared with currently available commercial receivers. The receivers at the monitor stations can track only 11 satellites at a time, and the tracking schedules cannot easily be revised or priority given as to which 11 satellites to track.[43] This results in a tracking gap of 3 to 4 hours per satellite per day. In addition, these receivers do not take full advantage of high-precision carrier phase data, which could be used to reduce multipath error contributions to the monitor station observables. Since some of the monitor sites suffer from very poor multipath environments, reduction of multipath errors is important. The main deficiency with current receivers, is that they can track only the Y-code and not the C/A-code, which is currently used by both civilians and the military. If there is a problem with the C/A-code, the MCS usually finds out only when C/A-code users call in to complain.

By upgrading the monitor stations with a high-quality, all-in-view receiver with C/A-code, Y-code, L_1, L_2, (and L_4) observables, OCS performance would be improved as follows: (1) the integrity of the C/A-code could be monitored, which would allow faster detection and correction of a problem by the OCS; (2) the high-precision carrier phase data could be used to reduce multipath error to the monitor station observables, thereby improving overall GPS accuracy; and (3) all satellites in view could be monitored, which would eliminate existing individual satellite tracking gaps of 3 to 4 hours per day and allow prioritized monitoring of any failing satellite signals.

Improvements to the monitor station facilities would require both software and hardware upgrades. Currently, the Air Force plans to award a $5 million contract to replace the monitor station receivers via a competitive bid in the summer of 1995. However, computer and software modifications required to take advantage of the improved receivers will not be upgraded at the same time. There also is a requirement in the 1995 OCS contract to replace the monitor station computers in order to take advantage of the new receivers, but there appears to be little coordination between the two procurements and little attention paid to the interfaces needed to optimize the system. The cost of replacing

---

[43] As many as 14 satellites can be in view of a monitor station at one time.

the monitor station computers and the software could not be obtained at the time of this report, since the contract had not been awarded.[44]

> ***Procurements for the replacement of the monitor station receivers, computers, and software should be carefully coordinated. The new receivers should be capable of tracking all satellites in view and providing C/A-code, Y-code, and L_1, and L_2 carrier observables to the OCS. Upgradability to track a new L_4 signal also should be considered. OCS software also should be made capable of processing this additional data.***

Backup Master Control Station. In view of the rapidly expanding use of GPS for both the military and civilians, it is critically important that the GPS be capable of continuous operation in all foreseeable contingencies. Currently, a considerable degree of redundancy exists in the space segment. However, very little if any redundancy exists in the operational control segment. Presently, a backup MCS is in place at the current OCS contractor's facility, but there are no firm, long-term plans to maintain such a facility. It is possible that the eventual implementation of the Block IIR autonomous navigation operation capability could remove some of the urgency for a backup system, but even so, such a capability will not be operational until near the year 2000 or later and will not completely eliminate the need for a backup MCS.[45] Air Force representatives have estimated that a backup MCS will cost around $14.4 million.[46]

> ***Firm plans should be made to ensure the continuous availability of a backup master control station.***

Operational Control Segment Simulator. Presently, there is no dedicated capability to test and prove out system hardware and software modifications or to train personnel in any new operational procedures resulting from the changes. Instead, the operational control segment and the space segment currently are used for testing and training purposes. This procedure not only imposes some degree of risk on the operational system and interferes with operational performance. Tests and training activities could be effectively performed in a facility that functionally simulates the operational system. This is a particularly critical issue in the near future because of the planned OCS upgrades and the deployment and

---

[44] Information provided by Capt. Earl Pilloud, Chief, GPS Control Segment, Air Force Space Command, 23 February 1995.

[45] Block IIR satellites have a military requirement to maintain a specified position accuracy for up to 180 days without clock and ephemeris updates from the MCS. This mode of operation is called autonomous navigation, or autonav. Autonav is accomplished by making inter-satellite pseudorange measurements using UHF (ultra high frequency) crosslinks and on-board processing to determine each satellite's ephemeris and clock offset.

[46] Memorandum from Col. Bruce M. Roang to the NRC committee, 23 December 1994.

operation of the Block IIR satellites. Also, if the recommendations of this report are implemented, a simulation facility would enable prompt and effective testing of the proposed modifications prior to their incorporation in the operational system. An Air Force estimate for the cost of an operational control segment simulator is $14.4 million.[47]

> *A simulator for the space and ground segment should be provided as soon as possible to test software and train personnel.*

Operational Control Segment Software. The current OCS system software was written several years ago. The hardware has since been upgraded, and over the years some software revisions have been made. However, the various upgrades have been written in different programming languages. This has produced a system that is lacking in modularity and is both difficult and expensive to maintain and upgrade. Because of this, an increasingly large percentage of the OCS budget is used to make relatively small changes to the system.

Since the original software was designed, significant improvements have been made in software development and management technology. Today, a system can be designed and implemented that would have improved reliability, longevity, and ease of enhancements through modular software engineering practice. Given the current state of the OCS software, the DOD's planned changes, and the recommendations contained in this report, the most economical and effective solution to this problem is to develop a new OCS software suite using current technology and methods. There is an option in the 1995 OCS upgrade procurement to either upgrade the existing software or to replace it with improved software that is easier to maintain and upgrade, but the choice is left up to the winning contractor.

> **The operational control segment software should be updated using modern software engineering methods in order to permit easy and cost-effective updating of the system and to enhance system integrity. This should be specified in the 1995 OCS upgrade request for proposal.**

Planned Block IIR Operation

Currently, each Block II/IIA satellite is updated once a day from the OCS with clock and ephemeris corrections generated by the MCS's Kalman Filter. As a result of military requirements, each Block IIR satellite will have a Kalman Filter on board and will be able to autonomously determine clock and ephemeris corrections independent of the OCS.[48] By

---

[47] Memorandum from Col. Bruce M. Roang to the NRC committee, 23 December 1994.

[48] Block IIR satellites have a military requirement to maintain a specified position accuracy for up to 180 days without clock and ephemeris updates from the MCS. This mode of operation is called autonomous navigation, or autonav. Autonav is accomplished by making inter-satellite pseudorange measurements using UHF crosslinks and on-board processing to determine each satellite's ephemeris and clock offset.

exchanging clock and ephemeris information every 15 minutes via UHF communications crosslinks, which will connect each satellite in the constellation to all of the other satellites in view, each satellite will have knowledge of the ephemeris and clock information of all the satellites in the constellation. Based on the 15-minute ranging data exchanged, the Block IIR satellites can autonomously update the navigation message being broadcast to users.

The current plan for testing the autonomous ranging capability is initially to download the 15-minute ranging data from each satellite's Kalman Filter once per day to the OCS so that it can be compared with the ground-based data derived from the MCS's Kalman Filter. After successful testing of autonomous satellite ranging capability is completed, clock and ephemeris corrections will be determined with the on-board Kalman Filter, and the satellites will automatically update the navigation message every hour. However, even with autonomous generation of clock and ephemeris corrections, the Air Force plans to continue daily uploads of the satellites' clock offset relative to UTC.[49] After 24 hours, the combined clock and ephemeris error for the Block IIR satellite constellation is expected to be 1.9 meters (1σ).[50]

Suggested Improvements Using the Autonomous Ranging and Crosslink Communication Capability

Current plans call for the use of the Block IIR satellite crosslink capability only for specific commands related to SA and autonomous navigation. Further improvements in accuracy, system reliability, and integrity could be obtained by exploiting the satellite ranging data obtained during the 15-minute autonomous ranging cycles and by more effectively utilizing the communication crosslinks. These improvements are discussed below.

Accuracy Improvements by Incorporating Satellite Ranging Data into Ground Solution

Since the satellite ranging data will initially be sent to the OCS for comparison with the ground-based data, the space-based measurements also could be incorporated into the MCS's Kalman Filter. By uploading these integrated corrections to the satellites, an incremental improvement in accuracy can be achieved over the initial planned Block IIR operational procedure, where the satellites will be uploaded with only ground-based clock and ephemeris corrections.

When autonomous satellite ranging capability has been activated, further accuracy improvements could be achieved if the integrated corrections were sent to satellites at least once per day. Ideally, one satellite could be sent the corrected data every hour and the crosslinks could be used to relay the information to all the other satellites. These integrated

---

[49] Source: Input provided to the NRC committee by Capt. Christopher Shank and Capt. Earl Pilloud, Air Force Space Command, January and February 1995.

[50] Response from Martin Marietta Astro Space Division of Lockheed-Martin, 6 February 1995.

corrections, instead of corrections generated autonomously on the satellites, could be used to update the navigation message every hour. In order to operate in this manner, the data rate of the Block IIR UHF communication crosslinks may have to be modified. The exact improvement to the combined clock and ephemeris error is not known, because a complete analysis was not conducted. However, more frequent uploads of integrated space-based and ground-based clock and ephemeris information should result in errors no greater than 1.2 meters (1σ).

> ***The planned Block IIR operation should be reexamined and compared to the accuracy advantages gained by incorporating inter-satellite ranging data in the ground-based Kalman Filter and uploading data at some optimal time interval, such as every hour, to all GPS satellites.***

Satellite Health Monitoring to Improve System Reliability and Availability

Since the Block IIR satellites will have a UHF communications crosslink capability, satellite health monitoring could be implemented that could improve overall system reliability and availability. For example, if a satellite detected an anomalous on-board health reading, but was not in contact with a ground station, it could relay the information through the crosslinks, enabling another satellite that was in contact with a ground station at that instant to download the information. The MCS in turn, could upload commands to the failing satellite via the crosslinks. This would improve the reliability of each individual satellite by minimizing out-of-service time, thus improving the percentage of time that a full 24-satellite constellation would be available to users.

> ***Block IIR satellite communication crosslinks should be used to the extent possible with the existing crosslink data rate to support on-board satellite health monitoring for improved reliability and availability and in order to permit a more rapid response time by the operational control segment.***

Ground-Based Integrity Improvements

The Block IIR communications crosslinks also could be used to improve GPS signal integrity for all users. For example, if an anomalous pseudorange signal was detected at a monitoring station, the MCS could upload a command to the satellite broadcasting the anomalous signal by relaying this command through the crosslinks. The faulty satellite could be commanded to broadcast a code that could not be tracked by a user's receiver, and would therefore, be dropped from the users' positioning solution.

To use the crosslinks to improve GPS integrity for PPS and SPS users, the receivers at the monitor stations must be upgraded to monitor the C/A-code. The data rates on the crosslinks must be able to support commands sent from the MCS.

The Block IIR inter-satellite communications crosslinks should be used to relay integrity information determined through ground-based monitoring.

PERFORMANCE IMPROVEMENTS TO ENHANCE THE MILITARY USE OF GPS

From its very inception, the force enhancement capabilities of GPS for both U.S. and allied armed forces has been one of the system's most important capabilities. This remains true today, despite the fact that civilian and commercial use of the system has grown rapidly. The current DOD policy is to secure for both the United States and its allies the full accuracy of GPS by using the encrypted Y-code on both L_1 and L_2, while denying that accuracy via SA to a potential enemy who, like most civilians, will have the C/A-code available only on L_1. However, with the widespread use and proliferation of DGPS, the accuracy degradation produced by SA is routinely eliminated and, in many cases, civilians have access to more accurate signals than the military. As the cost of DGPS equipment decreases, differential technology and capability will proliferate. Differential systems will therefore become difficult to identify and render inoperative in a conflict situation. Furthermore, because adversarial forces are far less likely to be concerned with collateral damage, the 100-meter (2 drms) stand-alone accuracy of the SPS already poses a risk for our forces operating in a theater of war.

Earlier in this chapter the NRC committee recommended that the DOD concentrate future efforts towards the denial of GPS capability to an enemy by jamming the L_1 signal, the L_4 signal (if added), and other frequencies that may be employed by enemy forces to broadcast differential corrections. This strategy implies that U.S. forces must be properly equipped to operate in a high jamming environment generated by both U.S. military and enemy jammers. Based on this objective, the remainder of this section recommends several near-term technical enhancements to improve the overall performance of military GPS user equipment operating in the presence of spoofing, jamming, and interference. The greatest improvement in user equipment performance will result from the combined implementation of all five recommended enhancements in a single integrated system. Possible operational procedures that could be used prior to the availability of each recommended enhancement also are discussed.

Recommended Technical Improvements to Military User Equipment

Rapid, Direct Y-Code Acquisition

Current military receivers are designed to first acquire the more powerful C/A-code before handing over to the encrypted Y-code. Upon receiver restart, or following a loss of signal lock, a PPS receiver must go through acquisition in which a two-dimensional time-frequency search is carried out by trial correlations. With current receivers, this search conventionally is done serially, resulting in seconds to minutes of acquisition time for the C/A-code prior to Y-code hand-over, depending upon the amount of signal blockage experienced and the movement of the vehicle carrying the receiver. If the L_1 signal is jammed, the current receivers cannot acquire the C/A-code and as a result are denied access to the encrypted Y-code as well as dual-frequency ionospheric corrections. One receiver improvement that would enhance military access to the encrypted Y-code in a jamming environment would be the ability to acquire the Y-code rapidly without first acquiring the C/A-code. A method for improving L_2 ionospheric corrections in an L_1 jamming environment is addressed later in this chapter.

In order to obtain direct Y-code access, the signal acquisition processing capability of current PPS receivers must be improved through the use of massively parallel correlators, built using application-specific integrated circuit (ASIC) technology.[51] The technology is now available that would allow the incorporation of at least 1,000 parallel correlators per receiver at a reasonable cost. This would allow direct Y-code acquisition within 2 seconds in a non-jamming environment, without prior acquisition of the C/A-code. This would allow faster receiver time-to-first-fix after power-down and would enhance signal availability after a blackout interval.[52] The ability to directly acquire the Y-code on L_2 would ensure that the selective denial of the L_1 signal and the C/A-code through spoofing and jamming would eliminate or seriously degrade an enemy's use of GPS without impacting U.S. capabilities. According to experts in the field of military receiver technology, the technology for direct Y-code acquisition is in hand and in fact, the current military "Plugger" receivers do try to directly reacquire the Y-code after signal loss.[53] A military receiver with the capability to initially acquire the Y-code directly could be developed in 9-15 months depending on: (1) the amount of input received from the military regarding specifications; (2) the level of trade-off accepted between jamming-to-signal ratio versus the amount of time for direct Y-code acquisition; and (3) the ASIC development.[54]

> *The development of receivers that can rapidly lock onto the Y-coded signals in the absence of the C/A-code should be completed. The deployment of direct Y-code receivers should be given high priority by the DOD.*

---

[51] Massively parallel correlators using ASIC technology, permit the receiver to compare, at very fast speeds, the internally generated pseudorandom noise codes to the received codes, which contain data about the satellite's position and time the code was transmitted.

[52] See Appendix K for calculations showing a direct Y-code acquisition time of 2 seconds with current ASIC technology.

[53] Personal conversation with Mr. Tyler Trickey, Rockwell-Collins, February 1995.

[54] Information provided to the NRC committee by Mr. Charles Trimble of Trimble Navigation Ltd., 31 March 1995.

Antenna Subsystem Improvements

The jamming analysis shown in Appendix L assessed three candidate techniques for improved jamming resistance: aiding of tracking loops with inertial sensors, increased processing gain via wider signal bandwidth, and nulling/directive antenna systems. Nulling antennas were found to provide the single biggest improvement in jamming performance — on the order of 25-35 dB.

Both military and civilian users have deployed multi-element antenna structures for several years. In the late 1970s, work on the multi-element military AE-1 began. This antenna system was designed to effectively null a single jammer. In both this unit and its derivative design, the AE-1A, element phase shifting and combining is carried out in the radio frequency analog domain. These units have been deployed on a number of aircraft, but have not yet been widely utilized on other military weapons systems, primarily due to cost and size considerations.

Recently, more effective techniques have been developed wherein element phase shifting and amplitude weighting is done after spread-spectrum signal correlation, eliminating the RF phase shift components in favor of lower-cost correlator ASIC's and signal processing at baseband. With processing gain applied before nulling and beam forming algorithms, much improved jamming-to-signal margins are available.[55] These and future developments aimed at reducing the size and cost of antenna structures should be actively pursued.

> ***Nulling antennas and antenna electronics should be employed whenever feasible
> and cost effective. Research and development focused on reducing the size and
> cost of this hardware should actively be supported.***

Inertial Aiding Improvements

Aiding can be defined as the usage of any non-GPS-derived user dynamics and clock information in GPS receiver signal-tracking and navigation functions. The availability of such data can have a profound impact on GPS signal acquisition time, code and carrier tracking thresholds, interference and jamming resistance, anti-spoofing capability, and receiver integrity. Aiding works by providing auxiliary observations, which sense a vehicle's motion parameters. Inertial aiding is especially effective because of its invulnerability to electromagnetic interference and because its error characteristics are complementary to those of radionavigation systems, that is, inertial noise errors are low frequency and GPS signal tracking errors are high frequency.

Since the earliest days of GPS, the military has exploited synergism, at first with loosely-integrated inertial navigation systems (INS)/GPS built around existing aircraft INS mechanization, and more recently with "embedded" INS in which the inertial sensor and

---

[55] These concepts have been privately developed and patented by the Magnavox Electronic Systems Company (MESC), Patent 4734701. MESC has been continuing to enhance these concepts since inception.

GPS components reside in the same box. Embedded architectures combine GPS tracking-loop estimates with INS accelerometer and gyroscope outputs to correct INS biases. This provides fast GPS loop-aiding commands for a 10 dB-15 dB increase in tracking threshold and jamming margins and also supports rapid pull-in after signal blockage. This is referred to as a "tightly-coupled" INS/GPS structure. In less sophisticated aiding systems, often referred to as "loosely-coupled" structures, inertial positions and GPS positions or pseudorange data are merged in a cascaded filter structure, missing the benefits of improved GPS signal tracking margins. However, these loosely-coupled INS/GPS structures do extend the length of time that inertial operations can provide useful accuracy and a GPS integrity check, and also speed GPS signal acquisition.

Historically, inertial aiding had been too expensive for many tactical military applications. It was not until the 1980s that less-costly strapdown ring-laser gyroscope technology became common aboard military aircraft. However, in the last 5 years there have been other encouraging developments that could lead to wider implementation of aided GPS in tactical military applications. Fiber-optic gyroscopes and solid-state accelerometer configurations have come into use, and more recently, batch-fabricated quartz rate sensors and quartz and silicon accelerometers have been developed. These technologies should have a major impact on the cost of aided receiver systems.

> *The development of low-cost, solid-state, tightly-coupled integrated inertial navigation system/GPS receivers to improve immunity to jamming and spoofing should be accelerated.*

Signal Processing Improvements

The estimation of path delay and Doppler for all satellites in view is the most fundamental task of any navigation receiver.[56] Conventionally, delay and Doppler parameter estimates are extracted in delay-lock and phase-lock tracking loops consisting of dedicated loop software and correlator ASIC channels for each satellite. The resulting pseudorange and carrier phase quantities are then fed to the navigation filter routines, wherein these estimates are combined to produce updated position and velocity. In this traditional setup, predating the availability of fast and inexpensive digital signal processing and reduced instruction set computing (RISC) capable of hundreds of millions of double-precision floating point instructions per second, raw correlator data from a given satellite are processed without reference to state and error from other tracking loops, or from the receiver as a whole.

Fast computing permits statistically optimal validation and weighting of correlator data from all satellites early in the processing chain, based upon the full receiver state model. By taking advantage of inter-satellite path correlations, and by rapidly adapting filter gains to encountered signal amplitude and noise fluctuations, tracking thresholds can be improved on the order of 10 dB, and tracking can be made resistant to spoofing, multipath

---

[56] Doppler refers to the relative shift of frequency due to satellite-to-user motion.

capture, and cycle slip. This signal processing approach also can be combined with the inertial aiding techniques above, whereby correlator data, as well as accelerometer and gyroscope data are combined in an optimal fashion.

> *The development and operational use of GPS receivers with improved integration of signal processing and navigation functions for enhanced performance in jamming and spoofing should be accelerated.*

Improved L_2 Ionospheric Correction

In a hostile environment where the L_1 signal is jammed, PPS receivers will have access only to L_2 signals, thus eliminating their ability to perform dual-frequency ionospheric corrections. In such situations, military users must rely on an ionospheric model such as that contained in the broadcast navigation message or on a single-frequency correction process such as DRVID (Difference Range Versus Integrated Doppler).[57]

As the GPS signal travels through the ionosphere, the modulating codes (Y and C/A) are delayed by an amount proportional to the inverse square of the frequency. To first order, the carrier phase is advanced by the same amount, producing an effect termed "code-carrier divergence." DRVID is a technique that exploits this effect to compute the change in ionospheric delay over time. In order to determine the total ionospheric delay, an initial delay value must be known. This would work well in a scenario in which a receiver is initialized in a clear environment, that is, outside the region of L_1 jamming, and then tracks signals into the jammed region using DRVID to make ionospheric corrections. The primary disadvantage of DRVID is that it relies on continuous carrier tracking, which is not likely to be possible in a high-jamming and possibly high-blockage environment.

Currently, single-frequency receivers employ the 8-parameter Klobuchar model that is contained in the broadcast navigation message. This model is considered to be effective in eliminating approximately 50 percent of the total ionospheric delay with a day-to-day variability of 20 percent to 30 percent. It is suggested that enhancements to this model could improve the performance to the 70 percent to 80 percent level.[58] Furthermore, based on the current performance of local-area DGPS, the NRC committee believes that local-area estimates of ionospheric conditions made just outside the jamming region could provide even greater improvements.

---

[57] P.F. MacDoran, "A First-Principles Derivation of the Difference Range Versus Integrated Doppler (DRVID) Charged-Particle Calibration Method," *JPL Space Programs Summary 37-62* II, 31 March 1970.

[58] J.A. Klobuchar, "Potential Improvements to the GPS Ionospheric Algorithm." Presentation at the GPS/PAWG Meeting, 14 July 1993, Peterson Air Force Base, Colorado.

Military receivers should be developed that compensate for ionospheric errors when L_1 is jammed, by improved software modeling and use of local-area ionospheric corrections.

Possible Interim Operational Procedures

The NRC committee believes that the most significant shortcoming of a GPS denial strategy is the current inability to operate in high levels of enemy jamming, while at the same time denying GPS to an adversary. The implementation of technical enhancements to military user equipment, such as direct Y-code acquisition capability, improved nulling antennas, better inertial aiding capabilities, enhanced signal processors, and improved L_2 ionospheric corrections would assist in the optimal solution to this problem. Although the NRC committee believes that these technical capabilities are now available, unfortunately, such capabilities are not currently fielded by the military.

GPS receivers are especially vulnerable during their signal acquisition phase. This weakness is magnified by the inability of most military GPS receivers to acquire the Y-code during periods when the C/A-code is being jammed. Future receivers capable of direct Y-code acquisition will go a long way toward solving this problem. In any event, tactics must be developed and put in place to facilitate acquisition during jamming. Some possible disciplines that can be implemented in the near-term are presented below.

(1) **Develop military procedures to remove jammers and DGPS stations.** As with existing plans to destroy radars in a hostile area, plans and procedures should be developed to remove jammers and DGPS stations.

(2) **Acquire the Y-code outside the jamming area.** Prior to entering the jamming area, the C/A-code can be used to acquire the Y-code. Once the Y-code is obtained, and while still within the active jamming area, PPS receivers should be operated continuously or be repowered every few hours in order to maintain accurate time. Accurate time will aid in faster, direct reacquisition of the Y-code. This technique can be extended to aircraft-based GPS-guided munitions using low-powered C/A-code retransmissions aboard, or by hard-wiring of time-transfer circuits.

(3) **Review training exercises, procedures, and policy manuals.** The current training procedures and policy manuals should be examined to make sure U.S. troops are properly instructed to operate in both hostile jamming and denial jamming environments. For example, ground forces can make use of natural terrain and man-made obstructions to obtain some shielding from ground-based jammers.

(4) **Schedule denial jamming/spoofing.** Tactically, the U.S. military can interrupt denial jamming/spoofing for short time periods, typically 2 to 3 minutes hourly, to assist those friendly forces in need of C/A-code to reacquire the Y-code. These scheduled times would be short and random to prevent hostile troops from taking advantage of interrupted jamming. Dependence upon this technique will diminish as improved training procedures and time discipline techniques are disseminated into the force structure.

(5) **Develop and utilize C/A-code selective denial techniques that minimize impact upon friendly L_1-only military receivers, such as the Plugger.** The L_1 selective denial analysis of Appendix J suggests a four-part approach to selective denial of C/A-code on the L_1 band:

- apply shaped denial jamming combined with spoofing;
- use a switchable retrofit bandstop RF front-end filter; and,
- improve clock discipline, through operator training.

IMPROVEMENT IMPLEMENTATION STRATEGY

Because of the relatively long life time of GPS satellites (5 to 10 years) and the length of time required to replace the total constellation of 24 satellites, opportunities for introducing enhancements and technology improvements to the system are limited.

Figure 3-6 shows the current plan for satellite replacements. According to the GPS Joint Program Office, current plans for the Block IIF contract include 6 short-term, and 45 long-term "sustainment" satellites. As currently planned, the Block IIF satellites will be designed to essentially the same specifications as the Block IIR satellites. The current program and schedule make it possible for another country to put up a technically superior system that uses currently available technology before the United States can do so. Under current planning, the earliest opportunity for an infusion of new technology in the GPS space segment would be after Block IIF, probably sometime after the year 2020.

As noted throughout this chapter, the NRC committee believes that there are significant improvements that could be made to the system that would not only enhance its performance for civilian and military use, but also make it more acceptable and competitive internationally. One method to incorporate technology in an efficient and timely manner is through a preplanned product improvement (P^3I) process. With this type of approach, planned changes and improvements could intentionally be designed into the production of the satellites at specific time intervals.

Assuming that the first improvements suggested in this report are incorporated in the later half of the Block IIR satellites, additional funding might be required to incorporate changes for the already completed Block IIR satellites. However, the NRC committee believes that the timely improvement in system performance is adequate justification for the additional cost. Recommended improvements to the space segment and the operational control segment are summarized in Tables 3-13 and 3-14.

In addition to the specific recommendations given in this report, the NRC committee also discussed several enhancements that it believes have particular merit and should be seriously considered for future incorporation. These items are discussed in Chapter 4. Although a few enhancements could be included on the Block IIR spacecraft, especially if a P^{3}I program were implemented, most of the enhancements would have to be incorporated in the Block IIF spacecraft design.

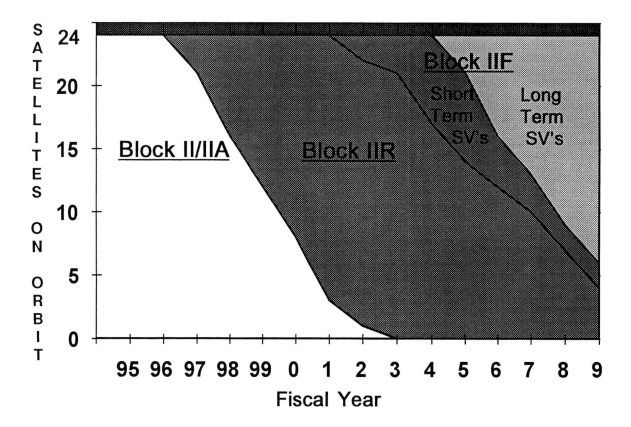

Figure 3-6 Current Plan for Satellite Replacement. (Courtesy of the GPS Joint Program Office)

Table 3-13 Space Segment Enhancements

Proposed Enhancement	Satellite Block	Implementation Date	Benefit of Enhancement
Turn SA to zero.	Block II/IIA, IIR, IIF	Immediately	Approximately 30-meter (2 drms) stand-alone accuracy for civil users.
Add a new L-band signal (would be usable before the Block IIR constellation is complete).	Block IIR, IIF	As soon as possible	Approximately 12-meter (2 drms) stand-alone accuracy for civil users. Enhanced integrity monitoring.
Use inter-satellite crosslinks to relay satellite health information and commands.	Block IIR, IIF	As soon as possible	Improve overall system reliability and availability.
Use inter-satellite crosslinks to relay ground-based integrity monitoring information and commands.	Block IIR, IIF	As soon as possible	Improve GPS signal integrity for all users.

Table 3-14 Operational Control Segment Enhancements

Proposed Enhancement	Implementation Date:	Benefit of Enhancement
Provide more frequent correction updates.	As soon as possible	Improve stand-alone GPS accuracy for PPS and SPS users (if SA is off and 48-hour embargo is lifted) by reducing combined clock and ephemeris errors by half.
Add more monitoring stations.	As soon as possible	Improve stand-alone GPS accuracy. Improve overall system reliability by allowing prompt detection of satellite anomalies. Allow for uninterrupted tracking of all satellites.
Improve Kalman Filter and dynamic models.	Added to 1995 Operational Control Station Request for Proposal	Improve accuracy by reducing combined clock and ephemeris errors with non-partitioned Kalman Filter (15%) and with improved dynamic model (5%).
Establish procurement coordination of improved monitor station receivers, computers, and software contracts.	As soon as possible and in conjunction with the 1995 contract award	Improve accuracy. Allow for integrity monitoring of C/A-code.
Reexamine planned Block IIR operation and compare to the accuracy advantages gained by incorporating inter-satellite ranging data in the ground-based Kalman Filter and uploading data at some optimal time.	Immediately	Possibly improve accuracy over planned Block IIR operation.

Proposed Enhancement	Implementation Date:	Benefit of Enhancement
Use Block IIR satellite communication crosslinks to the extent possible with the existing crosslink data rate to support on-board satellite health monitoring.	Block IIR satellites	Improved reliability and availability. Permits more rapid response time by the ground station.
Use Block IIR inter-satellite communication crosslinks to relay integrity information determined through ground-based monitoring.	Block IIR satellites	Permits more rapid response time for integrity monitoring.
Permanent backup master control station.	Immediately	Reduce risk and improve reliability of overall system.
Provide simulator to test software and train personnel.	Immediately	Reduce risk and improve reliability of overall system, improve efficiency of operations.
Update the operational control segment software using modern software engineering methods in order to permit easy and cost-effective updating of the system and to enhance system integrity. This should be specified in the 1995 OCS upgrade request for proposal.	Should be specified in 1995 Operational Control Station Request for Proposal.	Easier to make modifications to software. Reduces cost and complexity.

4

Technical Enhancements for Future Consideration

In the previous chapter, several GPS upgrades were proposed than could provide a stand-alone position accuracy approaching 5 meters (2 drms). Even though such an accuracy would satisfy many user requirements, as discussed in Chapter 2, better accuracy would still be required for many applications, such as Category II and Category III aviation landings, surveying and mapping, all-weather aircraft carrier landings, and some scientific applications.

Since satellite block changes are likely to occur at intervals of 5–10 years, there are a limited number of opportunities to take advantage of worthwhile technical advances and to refine the specifications based on new applications. Because of the anticipated worldwide dependance on the system, the committee believes that it would be shortsighted not to consider significant future improvements that would make GPS more generally useful and forestall the possible development of competing systems.

Below, several options for further GPS improvement are considered. Although the NRC committee determined that the supporting analyses for these options were not carried to the point where specific recommendations could be fully endorsed, the committee believes that the options presented have particular merit and should be seriously considered for future incorporation. Thus, these options are presented as suggestions for consideration rather than as recommendations. First, technical enhancements to improve the overall performance of the GPS for all users are presented; these are followed by enhancements that will benefit specific GPS user groups.

GPS IMPROVEMENTS TO IMPROVE OVERALL PERFORMANCE

Use of a 24-Satellite Ensemble Clock

Currently, clock offset corrections are determined on the ground and then sent to the individual satellites once a day as they pass over a GPS monitoring station. The Block IIR satellites will have the capability to determine their clock offsets autonomously relative to a space-based ensemble clock and exchange clock information with other satellites via crosslinks every 15 minutes. As a result, satellites can obtain clock information more often than once per day, which should result in a reduced clock error.

During autonomous ranging operation, each satellite will form an ensemble from the 14 satellites in view and will compare its offset relative to that ensemble. Further reduction of the clock error could be achieved if the clocks from all 24 satellites were used to create a single ensemble clock, as opposed to the current plan of letting each satellite form its own 14-satellite ensemble. For an ensemble of 14 clocks, the clock error is expected to be 1.1 meters (1σ) after a 4-hour period, as compared with an error of 0.9 meters (1σ) for a 24-satellite ensemble. This is discussed in greater detail in Appendix M.

The major advantage, however, of using a single, 24-satellite clock ensemble is not improved accuracy. Rather, it would allow quartz oscillators to be used on some satellites instead of atomic clocks, which are heavier, more expensive, require higher power, and have lower reliability than quartz clocks.[1] Since clock offset measurements are made frequently during autonomous ranging operation, the requirements on satellite oscillator stability are greatly reduced.[2] Therefore, quartz clocks could replace atomic clocks on at least some of the GPS satellites.[3] In addition, since atomic clocks require yearly maintenance, use of quartz clocks on some satellites also would reduce the ground control station workload.[4] Finally, the formation of an all-satellite ensemble clock may permit a failed clock in any one satellite to be detected and replaced more quickly and reliably.

In order to utilize an all-ensemble of all the 24 Block IIR GPS satellite clocks, the satellite software must be reprogrammed, and supporting ground software must be developed. In addition, further effort is needed to determine the optimal number and combination of quartz and atomic clocks.

Reduced Satellite Clock Errors Through Use of Improved Clocks

As discussed above, an ensemble reference clock can be used to reduce clock errors, relax requirements for clock stability, and eliminate the need for atomic clocks on some satellites. In order to improve the accuracy and the instantaneous frequency offset further, more accurate atomic clocks must be used on the satellites that will be carrying atomic

---

[1] According to Martin Marietta Astro Space Division of Lockheed-Martin, atomic clocks have been used in the past on GPS spacecraft and have provided a mixed heritage of superb stability and long life in some cases but unexplained premature degradation and failure in others. Each Block IIR satellite will carry two rubidium clocks and one cesium clock. The total cost of all three clocks represents approximately 3 percent of the price of the GPS spacecraft.

[2] With two-way time transfer measurements between satellites made every 15 minutes (900 seconds), the predictions need only to be good over this time period. Note that the clock error is the product of clock stability and prediction time. It is the reduction in prediction time from 1 day to 15 minutes that reduces the clock stability requirement by two orders of magnitude and, thus, enables the potential use of quartz oscillators.

[3] Since quartz clocks and atomic clocks have different frequency accuracies, their offsets would be weighted when determining a single ensemble time from all 24 satellites, that is, more weight would be given to the atomic clocks in the ensemble.

[4] Maintenance on the GPS clocks requires that each satellite is out of service 1 day per year.

clocks. A candidate future reference clock is the hydrogen maser. For terrestrial use, oscillators based on hydrogen masers have become the standard because they provide the best combination of low-phase noise, acceptable short- and long-term drift, reliability, and cost.[5] Hydrogen masers have been developed for space use, but none have been flown to date. It is possible that hydrogen masers could possibly be incorporated on Block IIF spacecraft, and the feasibility of doing so should be examined. If it appears viable, a research and development program should be initiated to develop a suitable space-qualified hydrogen maser oscillator suitable for GPS spacecraft.

Satellite-Based Integrity Monitoring

Perhaps the most innovative and promising method of signal integrity monitoring is through space-based monitoring, rather than ground-based monitoring. This capability known as Satellite Autonomous Integrity Monitoring or SAIM, would require the instrumentation of GPS satellites to monitor transmitted L-band signals from each other for accuracy and usability. If an anomalous signal is detected, neighboring satellites could inform the faulty satellite through the crosslinks. The faulty satellite could then autonomously begin broadcasting a code that could not be tracked by users' receivers. At the same time, the faulty satellite could inform the master control station (through the crosslinks) that there is a problem. With SAIM, the response time for commanding the faulty satellite to transmit a nonuseable code to the users after detection of a signal anomaly would be less than 1 second. Such a response time would meet many of the current integrity requirements, including those of the most stringent aviation applications.

In order to fully implement SAIM, however, extensive satellite modifications are necessary.[6] For example, a new crosslink design concept is required that is based on a CDMA (Code Division Multiple Access) protocol rather than the current Time Division Multiple Access (TDMA) protocol. This new crosslink would transmit the same navigation message observed by users to each neighboring satellite, which could then detect anomalies in the message. Since the required design modifications could be significant, fully operational SAIM would probably have to be incorporated in the Block IIF satellite design rather than the Block IIR satellites already under construction.

There is, however, a less extensive modification that could be incorporated in the Block IIR satellites to provide significant interim improvements in integrity monitoring. This modification would consist of the installation of a radio frequency field probe in the antenna near-field regions of the Block IIR satellite, which would monitor the integrity of its own satellite's L-band transmissions. Since the Block IIR crosslink transponder data unit is currently designed to transmit data every 36 seconds, the integrity information derived from

---

[5] Hydrogen masers are used for very long-baseline interferometry, which is used by Earth scientists to monitor tectonic deformations and Earth orientation.

[6] Based on information submitted to the committee by Martin Marietta Astro Space Division of Lockheed-Martin, which was reviewed by the Block IIR payload supplier, ITT, 24 January 1995.

the radio frequency probe could be transmitted to all other satellites in the constellation every 36 seconds. This would not be timely enough to meet many stringent integrity (time-to-alarm) requirements, but would provide much better integrity than is currently available.

Increased L_2 Signal Strength

One enhancement to the existing signal structure (C/A-code and Y-code on L_1, and the Y-code on L_2) that would improve performance for both civilian and military users is an increase in L_2 signal strength.

Currently, civilian receiver manufacturers attempt to correct for ionospheric errors through innovative codeless tracking techniques, with varying degrees of success. The chief limitation in the use of these somewhat expensive receivers is that, in order to function effectively, the carrier-to-noise ratio needed on the L_2 signal must be considerably higher than that required by a military PPS receiver. In many environments codeless receivers work very well. However, both L_2 pseudorange measurement precision and tracking margin for these receivers are considerably worse than for PPS receivers. In vehicle applications where there is signal attenuation due to foliage, the codeless receivers are more prone to signal loss. After loss of signal, the codeless receivers take a longer time to reach a given level of accuracy than a well-designed PPS receiver would.

In order to increase the L_2 signal strength by 6 dB, some spacecraft modifications must be made.[7] If an additional signal is added to the GPS satellites, as recommended in the previous chapter, then civilians would have access to another frequency for ionospheric corrections, so enhancements such as increasing the signal strength of L_2 would not be as vital. However, the military benefits obtained with an increased L_2 signal strength would not be addressed by adding another frequency, so such an enhancement should be considered on this basis alone.

As shown below, the performance of codeless receivers can be improved significantly if the transmitted power of the L_2 signal is increased by 6 dB.

Cross-Correlation Type Y-Codeless Receivers

This receiver recovers the L_2 observables by correlating the L_2 and L_1 Y-codes. A 6-dB increase in the GPS L_2 signal causes a 6-dB signal-to-noise ratio increase in the reconstructed L_2 carrier phase and pseudorange. This can be exploited in three ways:

---

[7] According to Martin Marietta Astro Space Division of Lockheed-Martin, an increase of 6 dB to 12 dB would require several spacecraft modifications. None are major except for DC power and thermal control, and these changes only become important at end-of-life when specification-to-performance margins will be lower than normally required on U.S. Air Force programs. Other factors such as harnessing, re-balancing, and panel re-layouts need to be assessed in detail but should not be significant problems. If an L_4 signal is also added to the Block IIR spacecraft, power sharing will be required, decreasing the amount by which the L_2 signal could be strengthened.

(1) The same antenna can be used. In this case the signals can be tracked to lower elevation angles. Given a typical gain fall-off from a survey-type antenna, the effective tracking limit would move from 15 degrees elevation to 5 degrees. Tracking to a 5-degree elevation angle is required for WAAS reference station sites and is beneficial for all differential GPS networks. Tracking to low elevation angles is also important when dual-frequency Y-codeless GPS is used on kinematic platforms such as aircraft, where bank angles can reduce antenna gain toward satellites at relatively high elevation angles.

In addition, reducing the minimum tracking angle from 15 degrees to 5 degrees will increase the maximum tropospheric signature by about a factor of three. For high-accuracy GPS users who solve for tropospheric delays either to remove it as an error source from baseline measurements or to monitor tropospheric parameters such as water vapor content, the lower elevation tracks give about a threefold increase in accuracy.

(2) In applications where the limiting error is signal multipath originating from reflectors at low elevation, the system designer may decide to exploit improved signal-to-noise ratio by specifying an antenna with more rejection at low-elevation angles.

(3) Under some conditions, ionospheric variations cause a Y-codeless receiver's L_2 tracking loop to slip cycles.[8] Given an L_2 signal with 6 dB more power, the receiver's L_2 tracking loop bandwidth could be increased by a factor of two.

L_2 Squaring Y-Codeless Receivers

This receiver recovers the L_2 observables by multiplying the L_2 Y-code by itself. A 6-dB increase in the GPS L_2 signal causes a 12-dB signal-to-noise ratio increase in the reconstructed L_2 carrier phase and pseudorange. These same benefits apply to the squaring receiver, with increased effects.

Y-Code Tracking PPS Receivers

For the military, a 6-dB increase in L_2 signal strength would assist in direct Y-code acquisition and would improve the anti-jam margin, especially if L_1 was jammed during a conflict. For example, if the power of the L_2 signal is increased by 6 dB, then 6 dB in anti-jam capability could be provided to military users. The important parameter is the ratio between the power of the desired signal and the jammer power. Since the latter decreases

---

[8] Personal communication between committee members and Bill Krabill, NASA, Wallops Island, March 1994.

as the square of the distance, a fourfold increase in signal power will allow the distance at which the jammer defeats GPS tracking to be halved.

MILITARY ENHANCEMENTS

Block IIF Signal Structure Military Enhancements

If the NRC committee's recommendation to add an additional, unencrypted L_4 signal on Block IIR satellites and increase the signal strength of L_2 is adopted, the one remaining area in which further improvement might be considered for Block IIF satellites is that of enhanced resistance to RF (radio frequency) interference for military users. To achieve this capability, wider-band signals than currently provided by the present Y-code can be used or the desired signals can be supplied at greater intensity, possibly using spot-beam techniques, which would illuminate an area of conflict.

Wide-Band Signals at High Frequency

A significant increase (approximately 10 dB) in anti-jam capability could possibly be achieved on the Block IIF satellites by using another wide-band signal, occupying perhaps 100 MHz to 200 MHz.[9] Such a broad signal would require that the carrier be at S-band frequency (approximately 2 GHz) or higher frequency. Although moving to a higher frequency would require receiver and spacecraft antennas to accommodate the signal, as well as other modifications, the move to a higher frequency would result in a reduced nulling antenna size and increase its performance. Such a high frequency would also provide increased immunity to the effects of ionospheric scintillation, which can degrade receiver performance when it is present.[10]

To demonstrate the effectiveness of a wide-band signal against a jammer (assumed to be co-located with a target), calculations have been performed for jammers operating at power levels of 100 watts and 10 kilowatts. (See Appendix L). At these two power levels, code and carrier tracking thresholds were estimated as a function of range from the jammer. For many applications, the key parameter is not the minimum range for loss of signal lock, but the minimum range for acceptable miss distance (range error) at the target. Therefore, the minimum range-to-jammer for a 1-meter range error was also determined.

---

[9] Additional information on wide-band signals is given in Appendix L.

[10] Ionospheric scintillation is a phenomenon in which the Earth's ionosphere introduces rapid phase and amplitude fluctuations in the received signals.

Figures 4-1 and 4-2 are the pseudorange errors as a function of distance for various receiver alternatives described in Appendix L and the two jammer power levels.[11] The difference between the narrower-band Y-code and wide-band options is rather dramatic, even on the log-log plots. The most capable system operates below the 1-meter level to within about 45 meters of the 100-watt source. At 1,000 meters, the code tracking error is below the centimeter level. As shown in Table 4-1, carrier phase tracking and code loop aiding are available within several hundred meters of the jammer. The miniaturized nulling antenna with aiding is good down to about 175 meters. Both wide-band options, which are combined with inertial aiding, are substantially more capable than the best performing existing Y-code system.

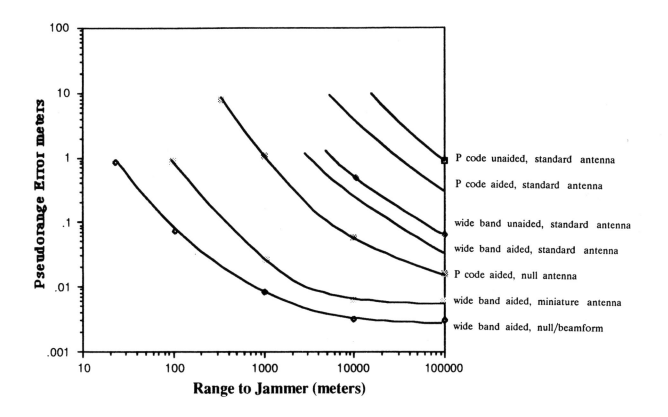

Figure 4-1 Wide-band GPS with a 100-watt jammer.

---

[11] Data generated by J. W. Sennott, Bradley University, Peoria, Illinois.

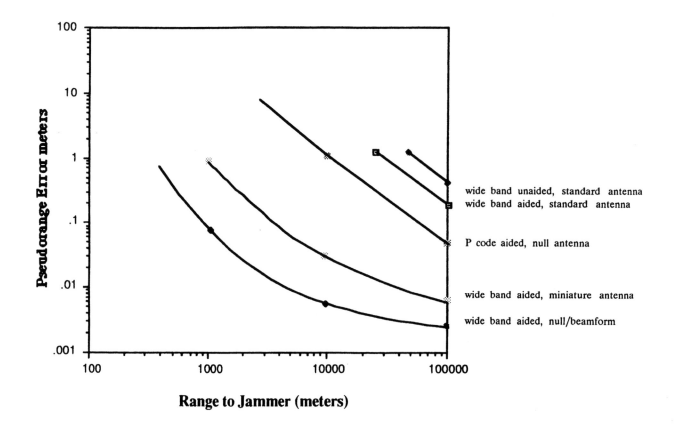

Figure 4-2 Wide-band GPS with a 10-kilowatt jammer.

    Tables 4-1 and 4-2 summarize the results of this exercise. The most significant finding, perhaps, is that with the wide-band signal using unaided tracking and a simple antenna, a vehicle can approach a 100-watt jammer to within 6 kilometers before a 1-meter range error has accumulated. With aided tracking, this range is reduced to about 3 kilometers. For many airborne weapons systems, this is sufficiently close to permit a successful mission when using inertial navigation for the balance of the flight, that is, assuming the worst case scenario in which the jammer and target are co-located. Considering that the size and cost of current nulling antennas may prohibit their use on certain weapon systems, this is a significant finding and supports the notion that consideration should be given to the eventual inclusion of a new, very wide-band waveform.

Table 4-1 GPS Wide-Band Signal Augmentation Performance with a 100-Watt Jammer

System Option	Code Status		Carrier Telemetry Status	
	Jammer distance at loss of lock (meters)	Jammer distance for 1-meter range error (meters)	Jammer distance at loss of lock (meters)	Range error at loss of lock (meters)
1. Y-code unaided standard antenna	18,000	90,000	90,000	1.0
2. Y-code aided standard antenna	10,000	35,000	21,000	---
3. Y-code aided nulling antenna	550	1,000	1,400	1.9
4. Wide-band unaided standard antenna	6,000	6,000	35,000	0.1
5. Wide-band aided standard antenna	3,100	3,100	6,500	0.27
6. Wide-band aided miniature antenna	175	175	450	0.19
7. Wide-band aided null/beamforming antenna	45	45	215	0.19

Table 4-2 GPS Wide-Band Signal Augmentation Performance with a 10-Kilowatt Jammer

System Scenario	Code Status		Carrier Telemetry Status	
	Jammer distance at loss of lock (meters)	**Jammer distance for 1-meter range error (meters)**	**Jammer distance at loss of lock (meters)**	**Range error at loss of lock (meters)**
1. Y-code unaided standard antenna	---	---	---	---
2. Y-code aided standard antenna	---	---	---	---
3. Y-code aided nulling antenna	---	20,000	---	---
4. Wide-band unaided standard antenna	---	60,000	---	---
5. Wide-band aided standard antenna	---	31,000	---	---
6. Wide-band aided miniature antenna	---	1,800	---	---
7. Wide-band aided null/beamforming antenna	---	450	---	---

Spot Beams

The advantages of introducing a new, 200-MHz wide-band signal at a higher carrier frequency for coping with a jamming environment were discussed above. While this offers the best technical solution, the difficulty of finding a suitable frequency band and the need to develop a new suite of military receivers to acquire the signal must be considered. An alternative solution to a wide-band signal for improved anti-jam margin would be the use of spot beams. By employing a steerable spot beam on the satellite to illuminate an area of conflict, the desired signal power at the receiver could be increased. For example, if a 3-meter, steerable-reflector L-band antenna (or phased-array antenna) could be added to the spacecraft, then a gain of approximately 20 dB would be obtained, which would increase range-to-jammer penetration by a factor of 10.[12] While adding a 3-meter steerable antenna to the GPS satellites is a very significant change with attendant complexity, weight, and cost penalties, this is clearly a preferable approach to simply boosting the overall L_2 transmitter power.

In summary, in addition to increasing the L_2 transmitted power, military anti-jam capabilities can be further improved by using a new, very wide-band signal (approximately 200 MHz), a spot beam, or some combination of both.

ENHANCEMENTS FOR HIGH-PRECISION USERS

GPS Transmit Antenna Calibration

High-accuracy users of GPS rely on differential carrier phase measurements to obtain millimeter- to centimeter-level results. High accuracies are obtained because for the differential measurements, most satellite-based errors are common mode errors and cancel in the differencing process. One error, however, that does not cancel is the error due to variations of the effective location (phase center) of the transmitting GPS antenna. These variations are a function of the angle to the user, primarily the angle off the GPS antenna array boresight.

Satellites that require precise orbit determination, such as Topex/Poseidon, are vulnerable to this error because the satellites view the GPS antenna from large angles off boresight. The maximum boresight angle to receivers on the ground is about 13 degrees, while the angle to a satellite in an orbit at 1,300 kilometers altitude is about 17 degrees. Variations of a few centimeters in the GPS transmit antenna phase center would induce variations of about 10 centimeters in the altitude of the Topex orbit. Even for ground-based measurements, these effects may contribute a small ($\sim 10^{-9}$ x baseline length) error. Phase variations are expected to be much greater at larger boresight angles. For such applications, knowledge of the transmit antenna phase variations is needed to reliably obtain centimeter, or subcentimeter, accuracy.

By measuring the transmit antenna phase center, the error currently limiting accuracy of very-high precision users can easily be eliminated. Since elaborate antenna measurements are already being made prior to launch, it should be relatively simple to make the measurements required to determine the actual phase center.[13]

---

[12] Since the nominal GPS antenna has a gain of +11 dBiC L_2 and +13 dBiC L_1, at 14.3 degrees off axis, the benefit of the postulated spot beam is about 20 dB.

[13] B. R. Schulper, B. L. Allshouse, and T. A. Clark, "Signal Characteristics of GPS User Antennas," *Navigation: The Journal of the Institute of Navigation* 41, no. 3 (1994).

Knowledge of Spacecraft Characteristics

Another limitation in the accuracy obtained by precision GPS users is caused by errors in the dynamic model used to predict the behavior of a GPS satellite as it expands and contracts due to the space environment. The current model could be improved if better information was available on the thermal emissivity of the exterior of the spacecraft surfaces, including its solar panels. In addition, telemetry information on temperatures of spacecraft surfaces is needed.

For long baseline applications, the limiting error source is usually the GPS orbit error, even when high-accuracy post-processed orbits are utilized. To reach accuracies of 10^{-9} times baseline length, the orbits must be known to about 10^{-9} times the distance to the satellites (approximately 20,000 km), or about 0.02 meters. In other terms, an unmodeled acceleration of 10^{-12} g would accumulate to 0.02 meters after a 12-hour orbit. Unequal radiation of heat from sides of the GPS satellite causes accelerations much larger than 10^{-12} g.

With a minimal amount of effort, the thermal properties of the materials on the exterior surfaces of the Block IIR satellite could be determined. To accomplish this, instruments to measure temperature could be added to the GPS satellites prior to launch, and the data received from these instruments could be transmitted to the ground. This would allow accelerations of the spacecraft surface, which result from uneven heating in space, to be calculated. These accelerations could then be included in models to improve accuracy. In the absence of deterministic models developed through actual measurements, these radiation pressure parameters can only be estimated.

Improved L₁ Signal Reception at Angles Below the Earth's Horizon

In order to support the Block IIR crosslink capability, the specifications related to the UHF antenna on the Block IIR satellites are different from the Block IIA specifications. As a result of these UHF antenna changes, the L-band antenna on the Block IIR satellites will be less symmetrical and will have a narrower pattern. For angles beyond the limb of the Earth as viewed from GPS satellites, this change will probably result in a reduction in the L_1 power currently observed with Block IIA spacecraft by approximately 3 dB.[14] For spacecraft applications of GPS, this reduction in received L_1 power over that currently observed with Block IIA satellites and the narrower antenna pattern could decrease the ability of low-Earth orbit and geosynchronous satellites to receive GPS signals and make GPS-based positioning more difficult from orbit. By increasing the L_1 signal power or improving the symmetry of the L-band antenna, spacecraft applications using GPS could be greatly enhanced.

---

[14] Currently, there is no official specification by the Air Force for end-of-life-power requirements beyond the Earth's horizon.

Appendix A

Study Participants

ACCQPOINT Communications Corp.
Art Humble

Aerospace Corporation
Mohan Ananda
Harold Bernstein
Chia-Chun (George) Chao
John Clark
Bill Feess
Paul Massatt
Prem Munjal
David Nelson
Bryant Winn, Jr.

Air Force Space Command
John Anton
Harrison Freer
Richard Koons
Earl Pilloud
Christopher Shank

Aircraft Owners and Pilots Association
Stephen Brown

Air Force Space and Missiles Center
Lynn Anderson
Kim Cirillo
Steve Decou
Charlie Golden
Bernard Gruber
Brian Knitt

Donald Latterman
Mark MacDonald
Ricardo Martinez
Al Mason
Charles Meadows
John Nauseef
Donna Shipton
Stephen Steiner
Earl Vaughn
Michael Wiedemer

Air Transport Association of America
William Russell

Allen Osborne Associates
Robert Snow

Alpha Instrumentation/Information Management (AI²M)
Daniel Alves

ANSER
David Martin

Association of American Railroads
Howard Moody

Ashtech Incorporated
Jonathan Ladd

Aviation Management Associates, Inc.
Larry Barnett

Booz•Allen & Hamilton
Michael Dyment

Central Intelligence Agency
Terry McGurn

Crown Communications
Harry Hodges

Deere & Company
Wayne Smith

Defense Intelligence Agency
Barry Joseph
Albert Glassman

Defense Mapping Agency
Mike Full
Steven Malys
Scott True
William Wooden

Department of Transportation
Peter Serini
George Wiggers

Differential Corrections Inc.
Ron Haley
Bruce Noel

Eagle/Lowrance Electronics
Steve Schneider
Rafi Kedar

E-Systems
Anton Gecan

Federal Aviation Administration
Paul Drouilhet
Joseph Dorfler
Dave Peterson
Martin Pozesky

Federal Highway Administration
Frank Mammano
Lee Simmons

Federal Highway Administration, Turner Fairbanks Highway Research Center
Jim Arnold

Garmin International
Gary Kelley

General Railway Signal
Gordon Quigley

Global Telecommunications & Information Systems
Eric Bobinsky

IVHS America
Jim Costantino
Jerry Marsh

INMARSAT
Jim Nagle

Interstate Electronics
Peter Canepa
Jim Grace

ITT Aerospace/Communications Division
Peter Brodie
Laurence Doyle
Thomas Ernst
Jon Schnabel

Jansky/Barmat Telecommunications, Inc.
Melvin Barmat

Jet Propulsion Laboratory
William Melbourne
George Purcell

John E. Chance & Associates, Inc.
Andy Bogle
Philip Stutes

**Johns Hopkins University,
Applied Physics Laboratory**
Lee Pryor

Joint Chiefs of Staff
Jim Burton
Pat Carlile
Joe Lortie
Tim Meyers
William Owens

Leica, Inc.
Paul Gaylean

Litton Aero Products
Abdul Tahir

Loral Federal Systems
Brian Hemley

Magellan Systems Corporation
Randy Hoffman
Jim White

Magnavox Electronic Systems Company
Walter Airth
Vito Calbi
Kenneth Lindenfelser

**Martin Marietta Astro Space Division of
Lockheed-Martin**
Robert Bebee
Aniruddha Das
Jim Graf
John Hrinkevich
John Mergen
Robbin Shultz

**Massachusetts Institute of Technology,
Lincoln Laboratory**
William Delaney
Jay Sklar

MITRE Corporation
Robert Berkowitz
Robert Bales
Bakry ElArini
Thomas Hsiao
Young Lee
Kan Sandoo
Andrew Zeitlin

Motorola
Robert Denaro

**National Aeronautics and Space
Administration, Wallops Island**
Bill Krabill

National Air Intelligence Center
Scott Feairheller
Jay Purvis
Frank Scenna

**National Institute of Standards and
Technology**
Marc Weiss

**National Oceanic and Atmospheric
Association**
Don Pryor
Benjamin Remondi

Odetics Precision Time Division
Don Greenspan

**Office of the Assistant Secretary of
Defense for Command, Control
Communications and Intelligence**
Noel Longuemare
Jules McNeff

**Office of the Assistant Secretary
of the Air Force for Acquisition**
> Matthew Brennen
> Lee Carrick
> Chad Pillsbury

PlanGraphics, Inc.
> Mike Kevanney

Rand Corporation
> Gerald Frost
> Irving Lachow
> Scott Pace

**Riverside County Flood Control and
Water Conservation District**
> William Young

Rockwell-Collins
> Tyler Trickey
> Mike Yakos

Rockwell International Corporation
> Rich Arris
> Denny Galvin
> Steve Fisher
> Steven Scott

RTCA, Inc.
> David Watrous

Sea River Maritime
> Robert Freeman

Scripps Institute of Oceanography
> Jean-Bernard Minster

Stanford University
> Bradford Parkinson

TASC
> Iris Roberts

**Tampa Bay Vessel Information and
Positioning System, Inc.**
> John Timmel
> Mike Shiro

3S Navigation
> James Danaher

Trifed Corporation
> Robert Ballew
> Louis Decker
> Ray Helmering

Trimble Navigation
> Ann Ciganer
> Walt Melton
> Charles Trimble

True Time
> Bruce M. Penrod

University Navstar Consortium
> Randolph Ware

U.S. Coast Guard
> George Privon

U.S. Department of Agriculture
> Galen Hart

U.S. GPS Industry Council
> Michael Swiek

Appendix B

Abbreviated Committee Biographies

LAURENCE J. ADAMS (NAE) is the retired President and Chief Operating Officer of the Martin Marietta Corporation. He joined Martin Marietta in 1948 after receiving a bachelors degree in aerospace engineering from the University of Minnesota. Mr. Adams has held over a dozen engineering, management, and senior leadership positions in the company, and was president of Martin Marietta Aerospace before becoming President and Chief Operationg Officer. He is an expert in many areas of space and missile engineering, including propulsion, materials structures and dynamics, safety, reliability, and systems effectiveness. Mr. Adams has been a member of United States Air Force committees and panels, and USAF Scientific Advisory Board studies and panels. Mr. Adams has served as chair of several NRC committees, including the Committee on Advanced Space Technology and the Panel on Small Spacecraft Technology, and is a former president of the American Institute of Aeronautics and Astronautics.

PENINA AXELRAD is an assistant professor in the Aerospace Engineering Sciences at the University of Colorado at Boulder. Prior to joining the faculty of the University of Colorado, she was a lecturer in the Department of Aeronautics and Astronautics at Stanford University, where she received her Ph.D. in 1991. Dr. Axelrad received her B.S and M.S. in aeronautical and astronautical engineering from the Massachusetts Institute of Technology. Her professional experience with GPS includes prior employment as a GPS program manager and lead systems engineer for Stanford Telecommunications and as a GPS consultant for various companies. Dr. Axelrad has published a number of papers in the GPS field and she is the 1994-1995 Western Region Vice President of the Institute of Navigation. She also is an associate editor of *NAVIGATION*, The Journal of the Institute of Navigation.

JOHN D. BOSSLER is the director of Center for Mapping at the Ohio State University and a professor in the Department of Geodetic Science and Surveying. Dr. Bossler was the Director of Charting and Geodetic Services at NOAA and is a retired Rear Admiral in the NOAA Commissioned Corps. Dr. Bossler is knowledgeable of GPS and has experience in ocean and land mapping, geodesy, global change research, land and ocean surveying, and high accuracy uses of GPS. Dr. Bossler is past president of AM/FM International, the American Congress on Surveying and Mapping, the Geodesy Section of the American Geophysical Union, and is president of the University Consortium of Geographic

Information Science. Dr. Bossler received his civil engineering degree from the University of Pittsburgh, and his M. S. and PhD in geodetic science from the Ohio State University. Dr. Bossler has served and chaired several NRC committees.

RONALD BRAFF is a Principal Engineer at the Center for Advanced Aviation System Development (CAASD) at The MITRE Corporation. Mr. Braff is an expert in navigation technology, a technical advisor for the FAA concerning the application of GPS in the National Airspace System, and the test director for the FAA's Local Area Augmentation System (LAAS) for GPS. While at MITRE, his past activities included management and technical contributions in the following areas for the FAA: applications of satellites to communications, navigation, and surveillance, operational research of the FAA's field maintenance system, and analysis of air traffic control automation. Mr. Braff is the editor of the peer reviewed quarterly, *NAVIGATION*, The Journal of The Institute of Navigation. He recently served on the NRC's Committee on Advances in Navigation and Piloting.

A. RAY CHAMBERLAIN has been Vice President of the American Trucking Associations, Inc. since 1984. In 1987, Dr. Chamberlain was appointed as Executive Director of the State of Colorado Department of Highways and later its successor, the Colorado Department of Transportation. He has served one term as president of the American Association of State Highway and Transportation Officials; and has served as chair of the National Research Council's Transportation Research Board and the National Association of State University and Land Grant Colleges. He has also served as Chief Executive Officer of Chemagnetics, Inc.; Executive Vice President of Simons, Li & Associates, Inc.; and President of Mitchell & Co., Inc. From 1969 to 1980, he was President of Colorado State University, where he held a variety of positions, including Dean of Engineering, Executive Vice President and Treasurer of the Governing Board. He is a member of the American Society of Civil Engineers. Dr. Chamberlain is on the Board of Directors for, Fort Collins Chamber of Commerce, the Food Production Foundation, and Synergetics International. He has served on several NRC committees and chaired the NRC's Transportation Research Board's Strategic Transportation Research Study on Highway Safety. Dr. Chamberlain obtained his B.S. in engineering from Michigan State University and his Ph.D. in engineering from Colorado State University. Dr. Chamberlain possesses a broad knowledge of surface transportation issues, including state and local issues as well as the freight industry.

RUTH M. DAVIS (NAE) is President and CEO of the Pymatuning Group, Inc. in Arlington, Virginia and a member of the National Academy of Engineering. Her research interests include automation, electronics, computers, and energy. Dr. Davis received her Ph.D. in mathematics from the University of Maryland in 1955. She joined the David Taylor Model Basin in 1955 and was head of the Operations Research Division there from 1957 to 1961. She has worked for the National Library of Medicine, the National Bureau of Standards, and was Deputy Undersecretary for Research and Engineering for the Department of Defense and an Assistant Secretary in the Department of Energy. Since 1981, she has been President of the Pymatuning Group, and an Adjunct Professor in the School of Engineering at the University of Pittsburgh. Dr. Davis is currently the chairman of the Aerospace

Corporation, and is on the board of seven Fortune 500 Companies. She is also a member the NRC's Aeronautics and Space Engineering Board and the Naval Studies Board. She is serving on the Committee on the Space Station, and the Panel for the Cooperation on Applied Science and Technology Program. Dr. Davis has received the Department of Commerce Gold Medal and the Ada Augusta Lovelace Award.

JOHN V. EVANS (NAE) is President and Director of COMSAT Laboratories, which is the largest research center devoted entirely to satellite communications research. Prior to his current position, Dr. Evans was Assistant Director of the MIT Lincoln Laboratory. Dr. Evans is the co-editor of *Radar Astronomy* and has published over a hundred papers on the topics of radar reflection and high-power radar studies of the upper atmosphere and ionosphere. Dr. Evans has served on several NRC committees and chaired the Committee on Solar Terrestrial Research. Dr. Evans has served on the U. S. National Committee of the International Union of Radar Science since 1968. While he was chair in 1978, Dr. Evans led a delegation of over 150 U. S. scientists to the General Assembly in Helsinki, Finland.

JOHN S. FOSTER, Jr. (NAE) is the retired Vice President of the Science and Technology Department at TRW Inc. He joined TRW in 1973 as head of the company's energy research and development programs. Prior to his employment at TRW, Dr. Foster served in two Presidential Administrations as Director of Defense Research and Engineering (DDR&E) within the Department of Defense. In this position he instituted new policies and procedures for the management of technology and systems acquisition, and personally contributed to the successful development of many advanced defense systems, including GPS. Dr. Foster received a bachelor's degree in mathematics and physics at McGill University in Montreal, and earned a Ph.D. in physics from the University of California at Berkeley. He joined the staff of the Lawrence Berkeley National Laboratory while he was still a student, and helped to establish the Lawrence Livermore National Laboratory in 1952. He later served as Director of the Laboratory. Dr. Foster holds several patents, and is the author of many publications in the fields of high-energy physics, defense technology, and electronic systems. He has served on several NRC committees, including the Study of Presidentially Appointed Scientists and Engineers, and the Panel on the Impact of National Security Controls on International Technology Transfer.

EMANUEL J. FTHENAKIS is the retired Chief Executive Officer and Chairman of the Board of Fairchild Industries. Previously, he had the position of Executive Vice President in charge of the company's Communications, Electronics, and Space Group. Mr. Fthenakis joined Fairchild in 1971 as Director of Information Systems at the Space and Electronics Division, and was founder and Chief Executive of American Satellite Company during its formative years. A native of Greece, and a naturalized U.S. citizen, Mr. Fthenakis graduated from the National Polytechnic University of Greece and from Columbia University in New York. He was a member of the technical staff at Bell Laboratories and later joined General Electric Company's Space Division as Director of Engineering, where he was involved in the development of strategic reentry vehicles and other missile programs. Between 1962 and 1969, Mr. Fthenakis founded, organized, and directed the Ford Space Division and was responsible for developing the first U.S. military communications satellite system. In 1982 he received a presidential appointment to serve on the National Security Telecommunications Advisory Council.

J. FREEMAN GILBERT (NAS) is with the Institute of Geophysics and Planetary Physics at the University of California, San Diego. He received his Ph.D. in 1956 from the Massachusetts Institute of Technology, and is widely published in the field of geophysics. Dr. Gilbert has served on a number of NRC committees and has served as a board member for the Computer Sciences and Telecommunications board, and the Earth Sciences and Resources Board, which he currently chairs.

RALPH H. JACOBSON is the President and Chief Executive Officer of The Charles Stark Draper Laboratory. Prior to holding this position, Mr. Jacobson served in the U.S. Air Force for 31 years, and retired at the rank of Major General. His career included tours as a tactical airlift pilot, a project officer for the Titan-II inertial guidance system, and a number of assignments in the U.S. Space Program. As a Brigadier General, Mr. Jacobson was assigned to the Space Shuttle Program Office at NASA Headquarters, and later was the Air Staff Officer responsible for the budget of the Air Force's space program. His last position was Director of Special Projects within the Office of the Secretary of the Air Force. Mr. Jacobson received a B.S. from the U.S. Naval Academy, an M.S. in astronautics from the Air Force Institute of Technology, and an M.S. in business administration from the George Washington University. He is a member of several boards, committees, and advisory groups in the national security and aerospace fields, and is a former member of the NRC Committee on the Enhanced, Lower Cost Air Force Space Systems.

IRENE C. PEDEN (NAE) is a professor of electrical engineering at the University of Washington. She joined the faculty of the University of Washington in 1961 after receiving her Ph.D. from Stanford University, and after holding a number of professional positions in industry. From 1991 to 1993 Dr. Peden served in the National Science Foundation as the Director of the Electrical & Communications Systems Division and the Director of the Engineering Infrastructure Development Division. Her expertise includes electrical engineering and radio science, and she has published a number of professional papers on these subjects. Dr. Peden has served as a board member and chair for dozens of professional and honorary societies, and has served on several NRC committees.

KEITH D. MCDONALD is President of Sat Tech Systems and Technical Director for Navtech Seminars, Inc. Previously, Mr. McDonald directed the FAA's Aeronautical Satellite Division, and managed the satellite applications and technology program. He was also the Scientific Director of the DOD's Navigation Satellite Program during the formative stages of the GPS program. Mr. McDonald has been active in RTCA, preparing guidelines for the use of satellite systems in aviation, and has received the RTCA Citation for Outstanding Service. He also has received the Institute of Navigation's (ION) Norman P. Hays Award for outstanding contributions to the advancement of navigation, and served as the 1990 ION President.

JAMES W. SENNOTT is a professor of electrical and computer engineering at Bradley University. He is an expert in navigation and positioning systems, estimation theory, multiple access, spread-spectrum communications, image processing, software design and microprocessor architectures. In addition to his work at Bradley University, Dr. Sennott has worked with the Department of Transportation; NASA Goddard Space Flight Center; Caterpillar Tractor,Co.; SatTech Systems, Inc.; Delco Electronics; Interstate Electronics; Track Recorders; COMSAT Laboratory; and the MITRE Corporation. Dr. Sennott has been the principal investigator on contracts funded by the FAA; the U. S. Coast Guard; Caterpillar Tractor, Inc.; and the U. S. Maritime Administration. In his work for the U. S. Coast Guard, Dr. Sennott assisted in the development and application of GPS methods, including DGPS. Dr. Sennott received his B.S. in electrical engineering from the University of Delaware in 1963 and his M.S. and Ph.D. from Carnegie-Mellon University.

JOSEPH W. SPALDING is Project Manager of the Advanced GPS Project at the United States Coast Guard Research and Development Center in Groton, CT. Mr. Spalding has been conducting research in GPS and DGPS for nine years, and has published a dozen technical reports on these subjects. His current projects at the Research and Development Center include systems that measure the integrity of GPS and DGPS performance both on-board ships and as static monitors for the Coast Guard DGPS service and vessel attitude determination by using an array of GPS antenna/receiver combinations. Mr. Spalding holds a B.S. in electrical engineering from the State University of New York Maritime College and an M.S. in computer science from the University of New Haven. He is also a licensed Merchant Marine officer holding a rating of Third Mate of Oceans.

LAWRENCE E. YOUNG is a technical group supervisor developing high precision radiometric systems for geoscience and spacecraft applications at Caltech's Jet Propulsion Laboratory (JPL). The last twelve years of this work have concentrated on the development of high-accuracy GPS technology including digital receivers, multipath reduction, nanosecond-level clock synchronization, and the use of GPS for kinematic platforms and satellite applications. Dr. Young has published a number of papers related to GPS receiver and antenna research. He received a B.A. in physics from Johns Hopkins University, and a Ph.D. in nuclear physics from the State University of New York at Stony Brook.

Appendix C

Overview of the Global Positioning System and Current or Planned Augmentations

ORIGINS AND DEVELOPMENT OF THE NAVSTAR GPS PROGRAM

The navigation, positioning, and timing system that is known today as the Global Positioning System (GPS) is a combination of several satellite navigation systems and concepts developed by or for the DOD (Department of Defense). The predecessors to GPS include the following satellite systems: (1) Transit, an operational system developed for the U.S. Navy by the Johns Hopkins Applied Physics Laboratory that is still in use today[1]; (2) Timation, an experimental program developed for the Navy by the Naval Research Laboratory that demonstrated the ability to operate atomic clocks on board orbiting satellites and was used as a system concept for GPS;[2] and (3) Project 621B, an Air Force study program originated in 1964 by Aerospace Corporation and the Air Force's Space and Missile Organization.[3] In addition, a DOD Four Service Executive Steering Group was established in 1968 to investigate the development of a Defense Navigation Satellite System that would satisfy all of the DOD's satellite navigation requirements.

By 1972, the best characteristics of each of these four programs had coalesced to form the general system characteristics and initial design parameters for the system now known as the NAVSTAR Global Positioning System.[4] The system configuration and a request for developmental funding was submitted to the Director of Defense Research and Engineering, and the Air Force agreed to become the Executive Agent for this joint system.

---

[1] The Transit system was put into operation in 1964. To date, approximately 28 satellites have been launched, and although an 8-satellite constellation is still operating, the DOD plans to phase out its use by 1996. Source of Information: Personal conversation with Lee Pryor of the Applied Physics Laboratory, Johns Hopkins University, 24 January 1995.

[2] Three satellites were launched during the experimental Timation program.

[3] No satellites were actually launched as part of the 621B study program.

[4] Although the system is still officially known as the NAVSTAR Global Positioning System (GPS), the NAVSTAR name is rarely used. For the remainder of this appendix, and throughout the rest of the report, the system is simply referred to as GPS.

The GPS program was approved in 1973, and a Joint Program Office (JPO) located at the Air Force Space and Missile Organization in El Segundo, California was established.

From its inception, GPS was designed to meet the radionavigation requirements of all the military services and those of civilian users as well. On February 22, 1978, the Air Force began launching experimental GPS satellites, termed Block I satellites, on Atlas F launch vehicles. After the third satellite successfully achieved orbit, testing of the system's capabilities began at Yuma Testing Grounds, Arizona. Using a portable receiver mounted in a truck moving at 80 kilometers per hour, the Air Force showed that the desired positioning accuracy of 10 meters in two dimensions was easily achievable. After tests with the first three experimental satellites proved successful, eight additional Block I satellites were launched to complete the design and testing phase of the GPS program.[5] Although these satellites, designed and built by Rockwell International, were intended to have a 3-year life span, they achieved an average operational life of almost 7 years, and one of the Block I satellites was still operating as of the date of this report.

The next series of satellites, termed Block II, was designed to be fully operational. The first Block II satellite was launched aboard an Air Force Delta II rocket on February 14, 1989.[6] The current GPS constellation consists of 24 Block II/IIA operational satellites, and as previously mentioned, 1 Block I experimental satellite.

The GPS JPO has done an outstanding job of developing and testing the systems and equipment for GPS, as well as acquiring the hardware and software needed to deploy the system. This excellent effort was recognized in 1994 with the award of the Collier Trophy to the JPO and several of the major contractors involved in the GPS program.[7]

GPS POLICY, MANAGEMENT, AND OPERATIONS

Department of Defense

Responsibility for the day-to-day management of the GPS program and operation of the system continues to rest with the Department of Defense, and is carried out primarily

---

[5] Only 10 of the 11 satellites actually achieved orbit, due to a launch failure on December 18, 1981.

[6] The first Block II GPS satellite was originally scheduled for launch in January 1987 aboard the Space Transportation System (Space Shuttle). After the 1986 Challenger accident, the Air Force decided to use expendable launch vehicles instead. For more information see, *Satellite Acquisition: Global Positioning System Acquisition Changes After Challenger's Accident,* U.S. General Accounting Office, Washington D.C., September 1987.

[7] The National Aeronautics Association has awarded the Collier Trophy each year since 1912 for achievement in aeronautics and astronautics in America. The Aerospace Corporation, the Naval Research Laboratory, Rockwell International Corporation, and IBM Federal Systems received the 1994 award along with the Joint Program Office.

by the Air Force.[8] GPS research and development is managed by the Space and Missile Systems Center at Los Angeles Air Force Base. Testing and evaluation is conducted jointly by the Air Force Operational Test and Evaluation Center at Kirtland Air Force Base, New Mexico, and Air Force Space Command at Falcon Air Force Base, Colorado. Operations and maintenance also are managed by Air Force Space Command. Procurement and budgetary oversight for GPS are managed by Program Element Monitors within the space systems office of the Assistant Secretary of the Air Force for Acquisition. Through fiscal year 1994, the cumulative procurement budget for the space and ground control segments of the GPS is approximately $3.5 billion; and research, development, testing and evaluation spending totals approximately $3.7 billion.[9]

DOD policy for the GPS program is set by the Under Secretary of Defense for Acquisition and Technology, with the help of the DOD Positioning/Navigation Executive Committee. This committee receives input from all of the DOD commands, departments, and agencies, and coordinates with the Department of Transportation (DOT) Positioning/Navigation Executive Committee.

Department of Transportation

In response to a request from the DOD, and in order to meet the needs of civil GPS users, the DOT established the Civil GPS Service (CGS) in 1987. The CGS is operated and managed within the DOT by the Coast Guard and consists of the following: (1) the Navigation Information Service, which provides GPS status information to civilian users; (2) the Civil GPS Interface Committee, which provides a forum for exchanging technical information in the civil GPS community; and (3) the Civil PPS Program Office, which administers the program that gives qualified civil users access to the Precise Positioning Service (PPS) signal, used primarily by the U.S. and allied armed forces.

In May of 1993, the Secretary of Defense and the Secretary of Transportation agreed to examine the operational, technical, and institutional implications of increased civil use of GPS in order to satisfy both military and civilian needs. The resulting joint DOD/DOT task force concluded its work in December 1993 with the release of a report titled *The Global Positioning System: Management and Operation of a Dual Use System — A Report to*

---

[8] As with all other federally funded navigation systems, the ultimate decision-making authority over GPS operations, in peacetime and in wartime, is the National Command Authority, consisting of the President, or the Secretary of Defense with the approval of the President.

[9] These figures cover fiscal years 1974–1994, are in 1995 dollars, and have been provided by the GPS Joint Program Office. During this same period the military services have spent approximately $1.4 billion on the procurement of user equipment. The $10 billion figure that is often quoted for the total cost of GPS is based on total spending for all segments of the system through fiscal year 2002 consistent with current congressional direction.

the Secretaries of Defense and Transportation.[10] In response to management recommendations made in the report, the DOT has established a DOT Positioning/Navigation Executive Committee to interface directly with the DOD Positioning/Navigation Committee. The duties of the committee chair have been assigned to the Assistant Secretary for Transportation Policy who, along with the Under Secretary of Defense for Acquisition and Technology, will co-chair the newly formed joint DOD/DOT GPS Executive Board. This management structure is illustrated in Figure C-1. The DOT Positioning/Navigation Executive Committee and the Assistant Secretary for Transportation Policy will act as the focal point for GPS plans and policies developed by a number of DOT agencies involved in the use of GPS. These organizations include the U.S. Coast Guard, The Federal Aviation Administration (FAA), the Federal Highway Administration (FHWA), and the Federal Railroad Administration (FRA). The Executive Committee will also receive input from the Civil GPS Service Interface Committee.

The Federal Radionavigation Plan

The *Federal Radionavigation Plan* is the official source of planning and policy information for each radionavigation service provided by the U.S. government, including GPS. It is jointly developed by the DOD and the DOT, and is updated biennially.[11] The *Federal Radionavigation Plan* represents an attempt to provide users with the optimum mix of federally-provided radionavigation systems, and reflects both the DOD's responsibility for national security, and the DOT's responsibility for public safety and transportation economy. It was first released in 1980 to Congress in response to the International Maritime Satellite Telecommunications (Inmarsat) Act of 1978 (P.L. 95-564).[12]

---

[10] U.S. Department of Defense and U.S. Department of Transportation, *The Global Positioning System: Management and Operation of a Dual Use System — A Report to the Secretaries of Defense and Transportation*, Joint DOD/DOT Task Force, December 1993.

[11] U.S. Department of Transportation and U.S. Department of Defense, *1992 Federal Radionavigation Plan*, DOT-VNTSC-RSPA-92-2/DOD 4650.5 (Springfield, Virginia: National Technical Information Service, January 1993).

[12] Inmarsat is a 75 member-state cooperative organization operating a satellite system to provide telephone, telex, data, and facsimile services to the shipping, aviation, offshore, and land mobile industries.

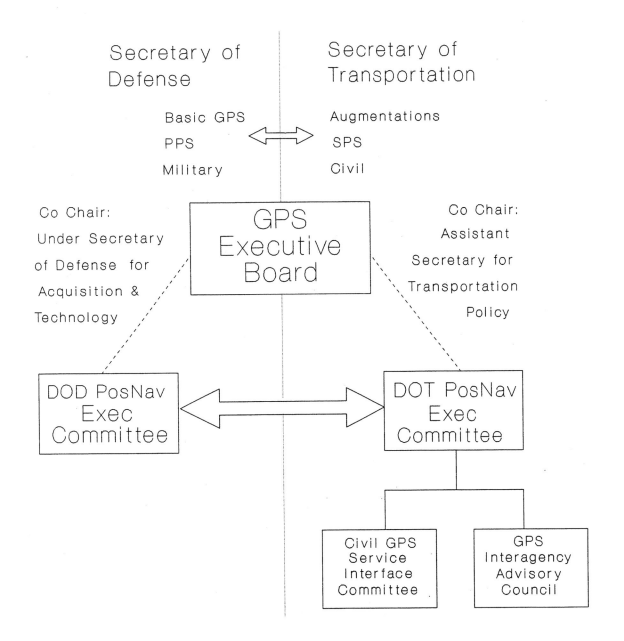

Figure C-1 GPS management structure as modified by *The Global Positioning System: Management and Operation of a Dual Use System — A Report to the Secretaries of Defense and Transportation.*

GPS TECHNICAL OVERVIEW

The technical and operational characteristics of GPS are organized into three distinct segments: the space segment, the operational control segment, and the user equipment segment. The GPS signals, which are broadcast by each satellite and carry data to both user equipment and the ground control facilities, link the segments together into one system. Figure C-2 briefly characterizes the signals and segments of the Global Positioning System, which are discussed in some detail below.

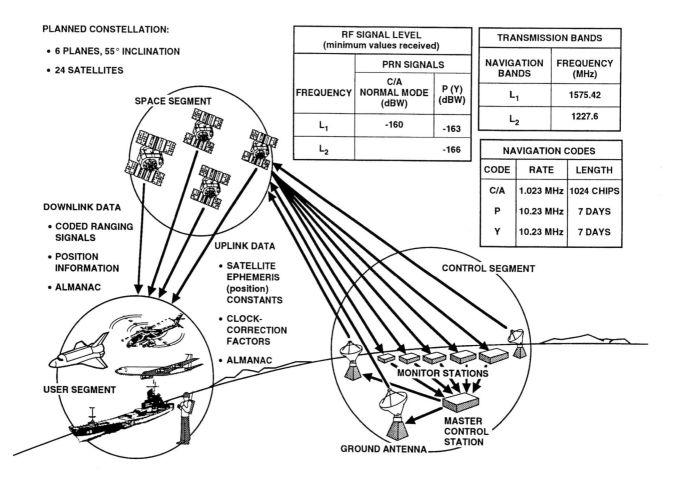

Figure C-2 Characterization of the GPS signals and segments. (Courtesy of the Aerospace Corporation)

Space Segment

The GPS constellation consists of 24 satellites, arranged in 6 orbital planes of 55-degree inclination, 20,051 kilometers (12,532 miles) above the Earth. Each satellite completes one orbit in one half of a sidereal day and therefore passes over the same location on earth once every sidereal day, or approximately 23 hours and 56 minutes. This particular orbital configuration and number of satellites allows a user at any location on the earth to have at least four satellites in view 24 hours per day. The constellation described above currently consists of 24 Block II/IIA satellites and one Block I satellite, which have been built for the U.S. Air Force by Rockwell International Satellite and Space Electronics Division, Seal Beach, California. Based on a fixed price, multi-year procurement contract totalling approximately $1.5 billion for 28 satellites, the unit cost of each satellite is approximately $53.8 million (1995 dollars).[13] Each Block II/IIA satellite is designed to operate for 7.5 years, but may operate beyond this life span based on the success of the Block I series. Figure C-3 shows a typical Block II/IIA GPS satellite.

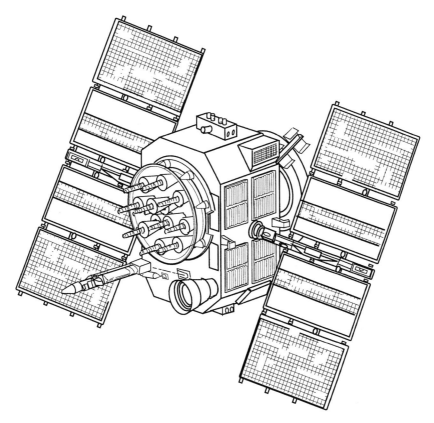

Figure C-3 Typical Block II/IIA satellite. (Courtesy of the Aerospace Corporation)

---

[13] U.S. General Accounting Office, *Satellite Acquisition: Global Positioning System Acquisition Changes After Challenger's Accident* (Washington, D.C.: U.S. Government Printing Office, September 1987), p. 11. Figures were converted to 1995 dollars using DOD Budget Authority inflation values for procurements.

Each Block II/IIA satellite weighs 1,881 kg (4,147 pounds) when fueled and is designed for a solo launch aboard an Air Force Delta II rocket.

The follow-on Block IIR replenishment satellite contract was competitively awarded in 1989 to Martin Marietta Astro Space Division, East Windsor, New Jersey for a total of 20 satellites. The estimated unit cost of each Block IIR satellite is $30.1 million (1995 dollars).[14] Recently, the Air Force exercised an option in the Block IIR contract to purchase one additional satellite. These satellites will also be carried into orbit by the Delta II rocket, with the first launch currently scheduled for 1996. Figure C-4 represents a typical Block IIR satellite.

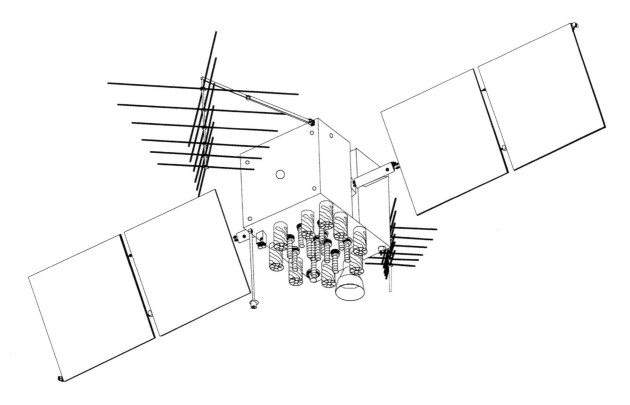

Figure C-4 Typical Block IIR satellite. (Courtesy of the Aerospace Corporation)

Although the Block IIR satellites are very different in appearance from the Block II/IIA satellites, they have been built to the same basic specifications and comprise the same kinds of components and subsystems. Many of the subsystems and components,

---

[14] U.S. General Accounting Office, *Airspace System: Emerging Technologies May Offer Alternatives to the Instrument Landing System* (Washington, D.C.: U.S. Government Printing Office, November 1992), p. 37. Figures were converted to 1995 dollars using DOD Budget Authority inflation values for procurements.

however, have been designed for improved performance and reliability, including the solar arrays, the gyroscopes, the batteries, and the nuclear-detonation detection system payload. In addition, the navigation payload on board the Block IIR satellites carries one cesium and two rubidium clocks, rather than the two rubidium and two cesium clocks present on the Block II/IIA spacecraft.[15] The Block IIR satellites also have two important operational capabilities not available from the Block II/IIA satellites. First, each subsystem and payload has been designed to allow on-orbit software reprogramming, allowing for much greater operational flexibility and upgrading, and second, the satellites can maintain specified positioning accuracy without contact from the operational control segment (OCS) on the ground for up to 180 days. This mode of operation, known as autonomous navigation or autonav, is accomplished by relaying positioning information between satellites using ultra-high frequency (UHF) inter-satellite links.[16]

The draft request-for-proposal (RFP) for the next generation of satellites beyond the Block IIR design, known as the Block IIF, is currently scheduled to be released in the spring of 1995, and the final version is currently scheduled for release in the summer. The first launch is anticipated in 2001.[17]

Operational Control Segment

The GPS operational control segment (OCS) consists of the master control station (MCS), located at Falcon Air Force Base in Colorado Springs, Colorado; remote monitor stations, located in Hawaii, Diego Garcia, Ascension Island, and Kwajalein; and uplink antennas located at three of the four remote monitor stations and at the MCS.[18] The four remote monitor stations contribute to satellite control by tracking each GPS satellite in orbit, monitoring its navigation signal, and relaying this information to the MCS. These four stations are able to track and monitor the whereabouts of each GPS satellite 20 to 21 hours per day. Land-based and space-based communications are used to connect the remote monitoring stations with the MCS.

The MCS is responsible for overall satellite command and control, which includes maintaining the exact orbits of each satellite and determining any timing errors that may be present in the highly accurate atomic clocks aboard each satellite. Errors in a satellite's orbital position or in a satellite's timing are determined by analyzing the same signal and

---

[15] Detailed information about the GPS Block IIR rubidium frequency standards can be found in: William J. Riley, "Rubidium Atomic Frequency Standards for GPS Block IIR," in *Proceedings of ION GPS-92, 5th International Meeting of the Satellite Division of the Institute of Navigation* (16-18 September 1992).

[16] Because Block IIR satellites will be launched on need to replace failing Block II/IIA satellites, it is impossible to determine exactly when autonav capability will become operational.

[17] According to the GPS Joint Program Office, current plans for the Block IIF contract include 6 short-term, and 45 long-term, "sustainment" satellites.

[18] A backup MCS also exists at Loral Federal Systems in Gaithersburg, Maryland.

navigation message from each satellite that is used by GPS receiver equipment. Using a Kalman Filter, computers at the MCS process the data collected at all the monitor stations in order to estimate these errors.[19] Updated orbits and clock corrections are relayed once a day to each satellite by the four ground antennas.

The day-to-day operations at the MCS are carried out by personnel belonging to the 2nd Space Operations Wing of Air Force Space Command. Routine maintenance is also conducted by the Air Force and its contractors. Remote monitoring stations are largely automated, but a small number of contract personnel do monitor and maintain each station's equipment. Average annual personnel and maintenance cost for the MCS, the four remote monitoring stations, and all their associated equipment is approximately $30 million.[20]

User Equipment

GPS user equipment varies widely in cost and complexity, depending on the receiver design and application. Receiver sets, which currently vary in price from approximately $400 or less to $30,000, can range from simple one-channel devices that only track one satellite at a time and provide only basic positioning information, to complex multi-channel units that track all satellites in view and perform a variety of functions. Most GPS receivers, however, consist of the same three basic components: (1) the antenna, which receives the GPS radio signal and in some cases provides anti-jamming capabilities; (2) the receiver-processor unit, which converts the radio signal to a useable navigation solution; and (3) a control/display unit, which displays the positioning information and provides an interface for receiver control.

The subsections of a typical GPS receiver-processor unit include the front-end section, the digital signal processor, and the microprocessor. The front-end section translates the frequency of a GPS signal arriving at the antenna into lower or intermediate frequency (IF) and converts the signal from analog to digital. This more manageable signal is then passed to the digital signal processor, which "tunes in" to these signals using tracking loops that compare incoming signal data to internally generated models of the satellite signals. GPS receivers normally track more than one signal at a time using multiple channels, but also can track multiple signals using either a single channel sequenced between satellite signals or a multiplexing channel. Once the digital signal processor is successfully tracking a set of GPS signals, the ranging data it extracts is passed to the microprocessor, where

---

[19] A Kalman Filter incorporates both observations and mathematical models of the system dynamics to produce an optimal estimate of the current state of a system. By using knowledge of how the system state can change over time, the Kalman Filter allows the contributions of individual measurement errors to be averaged. In the MCS filter, the system state includes satellite orbital parameters, clock parameters, and numerous other elements.

[20] U.S. Department of Defense and U.S. Department of Transportation, *The Global Positioning System: Management and Operation of a Dual Use System — A Report to the Secretaries of Defense and Transportation*

computer software converts it into information that can be usefully displayed for a user, such as position coordinates, or input to another type of user equipment, such as an inertial navigation system.[21]

Although the functions of a current GPS receiver are the same as those present in user equipment tested in the 1970s, they have little else in common. The size and cost of user equipment has decreased dramatically, while capabilities and the size of the commercial market continue to increase. In 1993, the total value of the GPS user equipment market was estimated to be $420 million, with over 100 companies marketing GPS receivers.[22] U.S. manufacturers maintain a competitive advantage over their Japanese counterparts, who are currently the principal competitors. However, the advantage could easily be lost. Larger U.S. companies, like Trimble Navigation, Ltd., invest as much as $25 million per year in GPS research to maintain their technological advantage. At the present time, U.S. domestic sales per unit represent less than 50 percent of the worldwide GPS market, and 45 percent of U.S. industry sales are to overseas markets.[23]

GPS Signal Characteristics and Operational Concepts

The GPS relies on the principle of "pseudoranging" to provide accurate positioning to its users. Each satellite in orbit continuously transmits a radio signal with a unique code, called a pseudorandom noise (PRN) code, that includes data about the satellite's position and the exact time that the coded transmission was initiated, as kept by the satellites' on-board atomic clocks. A pseudorange measurement is created by measuring the distance between a user's receiver and a satellite by subtracting the time the signal was sent by the satellite from the time it is received by the user.[24]

Once three ranges (or distances) from three known positions are measured, a position in all three dimensions can be determined. In the case of GPS, however, a fourth satellite is generally needed in order to eliminate a common bias in the pseudoranges to all satellites caused by a lack of synchronization between the satellite and receiver clocks. Once this clock bias is eliminated by the presence of a fourth signal, a highly accurate three-dimensional position can be determined. Figure C-5 below further illustrates the GPS pseudoranging concept.

---

[21] The coordinate reference system utilized by most GPS receivers is the World Geodetic System 1984 (WGS 84). WGS 84 is the fourth global geocentric coordinate system developed by the DOD since 1960.

[22] These estimates have been provided by the U.S. GPS Industry Council.

[23] Response from Trimble Navigation Limited, Sunnyvale, California, 13 September 1994.

[24] This measurement is also affected by signal delay caused by the Earth's atmosphere, as will be discussed later in this appendix.

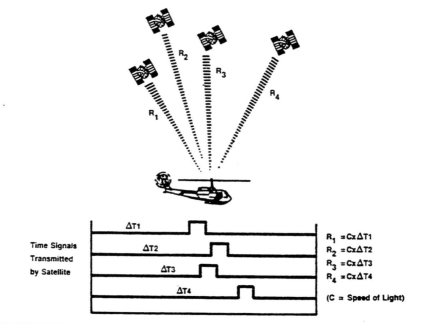

USER OBTAINS PSEUDO RANGE MEASUREMENTS
(R_1, R_2, R_3, R_4) TO 4 SATELLITES

Time Signals
Transmitted
by Satellite

$R_1 = C x \Delta T1$
$R_2 = C x \Delta T2$
$R_3 = C x \Delta T3$
$R_4 = C x \Delta T4$

(C = Speed of Light)

USER SET PERFORMS THE NAV SOLUTION FOR POSITION

PSEUDO RANGES:

$\boxed{R_1} = C \Delta t_1$

$\boxed{R_2} = C \Delta t_2$

$\boxed{R_3} = C \Delta t_3$

$\boxed{R_4} = C \Delta t_4$

POSITION EQUATIONS:

$(X_1 - U_X)^2 + (Y_1 - U_Y)^2 + (Z_1 - U_Z)^2 = (R_1 - C_B)^2$

$(X_2 - U_X)^2 + (Y_2 - U_Y)^2 + (Z_2 - U_Z)^2 = (R_2 - C_B)^2$

$(X_3 - U_X)^2 + (Y_3 - U_Y)^2 + (Z_3 - U_Z)^2 = (R_3 - C_B)^2$

$(X_4 - U_X)^2 + (Y_4 - U_Y)^2 + (Z_4 - U_Z)^2 = (R_4 - C_B)^2$

R_i = PSEUDO RANGE (i = 1,2,3,4)
  θ PSEUDO RANGE INCLUDES ACTUAL DISTANCE BETWEEN SATELLITE AND USER PLUS SV CLOCK BIAS.
    USER CLOCK BIAS, ATMOSPHERIC DELAYS, AND RECEIVER NOISE
  θ SV CLOCK BIAS AND ATMOSPHERIC DELAYS ARE COMPENSATED FOR BY INCORPORATION OF
    DETERMINISTIC CORRECTIONS PRIOR TO INCLUSION INTO NAV SOLUTION

X_i, Y_i, Z_i = SATELLITE POSITION (i = 1, 2, 3, 4)
  · • SATELLITE POSITION BRAODCAST IN NAVIGATION 50 Hz MESSAGE

RECEIVER SOLVES FOR:
  • U_x, U_y, U_z = USER POSITION
  • C_B = USER CLOCK BIAS

Figure C-5 Pseudorange concept. (Courtesy of the Aerospace Corporation)

Instead of transmitting one PRN code on one radio signal as described above, each satellite actually transmits two distinct spread spectrum signals that contain two different PRN codes, called the Coarse Acquisition (C/A) code and the Precision (P) code. The C/A-code is broadcast on the L-band carrier signal known as L_1, which is centered at 1575.42 MHz. The P-code is broadcast on the L_1 carrier in phase quadrature with the C/A carrier and on a second carrier frequency designated as L_2, that is centered at 1227.60 MHz. Figure C-6 illustrates the characteristics of both the L_1 and the L_2 signals.

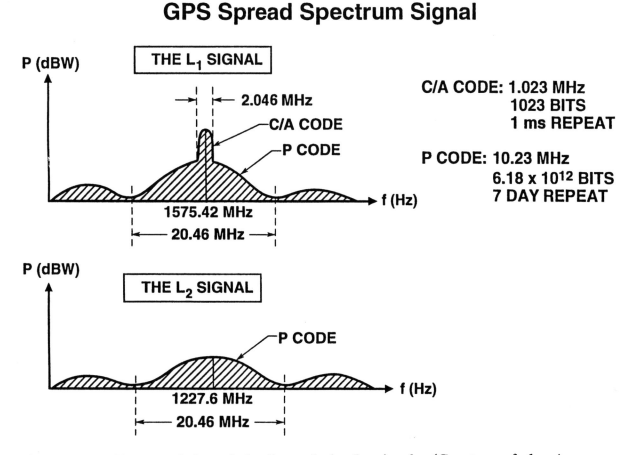

Figure C-6   Characteristics of the L_1 and the L_2 signals. (Courtesy of the Aerospace Corporation)

The L_1 C/A-code provides free positioning and timing information to civilian users all over the world, and is known as the Standard Positioning Service (SPS). The timing information on the C/A-code is also used by some receivers to aid the acquisition of the more accurate P-code. The P-code is normally encrypted using National Security Agency cryptographic techniques, and decryption capability is available only to the military and other authorized users as determined by the DOD. When encrypted, the P-code is normally referred to as the Y-code. The encryption process utilized, known as Anti-Spoofing (A-S), denies unauthorized access to the Y-code, and also significantly improves a receiver's ability to resist locking onto mimicked GPS signals, which could potentially provide incorrect positioning information to a GPS user.[25] Y-code availability through authorized decryption capability is known as the Precise Positioning Service (PPS).

Selective Availability and Other Positioning Errors

Before the PPS and SPS were established by the DOD with their current specified accuracy levels, the designers of GPS had anticipated that use of the Y- and C/A-codes would produce very different levels of positioning accuracy. Use of the Y-code was expected to result in 10-meter accuracy, whereas the C/A-code was expected to provide accuracy of 100 meters. Developmental testing of Block I GPS satellites, however, revealed that the accuracy difference between the two codes was not this significant. A report developed by the Joint Chiefs of Staff in the late 1970s highlighted this fact, and recommended that the GPS accuracy made available to civilians should be limited to 300 meters to 500 meters for national security reasons.[26] The precise positioning and standard positioning services were soon established, with PPS accuracy officially specified as 16 meters (SEP), and SPS accuracy specified as 500 meters (2 drms).[27] The SPS accuracy level was later changed to 100 meters (2 drms), as announced by the Under Secretary of Defense for Research and Engineering on June 28, 1983. This two-level accuracy arrangement is made possible on the Block II/IIA satellites through an accuracy denial method known as SA (Selective Availability), which was activated on March 25, 1990.

SA is a purposeful degradation in GPS navigation and timing accuracy that controls access to the system's full capabilities. SA is accomplished in part by intentionally varying the precise time of the clocks on board the satellites, which introduces errors into the GPS signal. This component of SA is known as dither. A second component of SA, known as epsilon, can also add error to the signal by providing incorrect orbital positioning data. PPS receivers with the appropriate encryption keys can eliminate the effects of SA. SA-induced errors can be varied by the DOD at the request of the National Command Authority or eliminated altogether, as was the case during the Persian Gulf War and the initial

---

[25] The process of sending incorrect information to an adversary's radio equipment (in this case a GPS receiver) without their knowledge, using mimicked signals, is known as spoofing.

[26] *Potential Military Exploitation of the NAVSTAR GPS by Adversary Nations* (Washington, D.C.: Organization of the Joint Chiefs of Staff, date unknown)

[27] SEP, or spherical error probable, represents an accuracy that is achievable 50 percent of the time in all three dimensions (latitude, longitude, and altitude). PPS accuracy is normally represented in this manner. SPS accuracy, however, is normally represented using a horizontal 2 drms measurement, or twice the root mean square radial distance error. Normally, 2 drms can be graphically represented as a circle about the true position containing approximately 95 percent of the position determinations. 2 drms, and other positioning accuracy definitions are discussed in greater detail in Appendix D.

operations in Haiti.[28] Figure C-7 displays the specified positioning, timing, and velocity accuracies for both the SPS and the PPS.[29]

PPS SPS	50TH PERCENTILE	drms	2 drms
POSITION			
• HORIZONTAL	8 m / 40 m	10.5 m / 50 m	21 m / 100 m
• VERTICAL	9 m / 47 m	14 m / 70 m	28 m / 140 m
• SPHERICAL	16 m / 76 m	18 m / 86 m	36 m / 172 m
VELOCITY			
• ANY AXIS	0.07 m/sec	0.1 m/sec	0.2 m/sec
TIME			
• GPS	17 nsec / 95 nsec	26 nsec / 140 nsec	52 nsec / 280 nsec
• UTC	68 nsec / 115 nsec	100 nsec / 170 nsec	200 nsec / 340 nsec

NOTES:
- FORMAL GPS SYSTEM ACCURACY SPECIFICATIONS ARE SHOWN IN THE SHADED AREAS.
- DERIVED GPS SYSTEM ACCURACY VALUES ARE SHOWN IN THE UNSHADED AREAS.
- THERE IS NO SPS USER VELOCITY ACCURACY SPECIFIED.
- 50TH PERCENTILE IS EQUIVALENT TO CEP, SEP, ETC.

Figure C-7 PPS and SPS specified accuracies. (Courtesy of the GPS JPO)

In practice, there are several additional sources of error other than selective availability that can affect the accuracy of a GPS-derived position. These include unintentional clock and ephemeris errors, errors due to atmospheric delays, multipath errors, errors due to receiver noise, and errors due to poor satellite geometry. Each of these error sources is discussed below and summarized in Table C-1.

---

[28] Selective Availability is normally on, but the level of error added to the GPS signal can be set to zero.

[29] It should be noted that these are specified accuracies not observed accuracies. Many GPS receivers currently in use are able to achieve better results than the specifications call for.

Table C-1 GPS Positioning Errors[a]

Error Source	Range Error Magnitude (meters, one sigma)	
	SPS	PPS
Selective Availability	24.0	0.0
Atmospheric Delay	7.0	0.7
Clock and Ephemeris	3.6	3.6
Multipath	1.2	1.8
Receiver Noise	0.6	0.6
Total User Equivalent Range Error (UERE)[b]	25.3	4.1
Typical Horizontal DOP (HDOP)[c]	2.0	2.0
Total Stand-Alone Horizontal Accuracy, 2 drms[d]	101.2	16.4

a.   The error budget figures included in this table are conservative estimates for a typical stand alone C/A-code receiver using standard correlation techniques, and a typical dual frequency Y-code receiver. This information was provided by the Jet Propulsion Laboratory (JPL) of the National Aeronautics and Space Administration (NASA), Pasadena, CA. Notes related to each component of this error budget, and the assumptions made to derive its value, are provided with Table 3-1 in Chapter 3.

b.   The total UERE is determined by adding the squares of the individual error magnitudes and taking the square root of the total.

c.   Dilution of precision (DOP) is discussed below, and HDOP is mathematically defined in Appendix D.

d.   The 2 drms horizontal positioning error is equal to 2 times UERE times HDOP. This mathematical relationship is further defined in Appendix D.

Atmospheric Error

Atmospheric error is caused by the delay of the GPS signal as it passes through the Earth's atmosphere. Part of this delay is due to the troposphere and part is due to the ionosphere. Because the ionospheric effect is dispersive and is a function of frequency, dual-frequency GPS receivers can calibrate this effect by measuring the differential delay and/or phase advance between the L_1 and L_2 frequencies, thus eliminating a great deal of the atmospheric error.

Civil users do not have direct access to dual frequency observations but have several means for reducing the ionospheric error contribution. For stand-alone navigation most C/A-code receivers apply an ionospheric correction, known as the Klobuchar Model, which can correct approximately 50 percent of the total ionospheric delay.[30] The model parameters are transmitted in the navigation message and are updated infrequently. High performance C/A-code receivers often perform codeless or cross-correlation tracking of the L_2 signal to permit them to derive ionospheric correction parameters. These techniques suffer from substantial signal-to-noise ratio losses and do not work well in high-blockage or high-dynamic situations.

Tropospheric delay cannot be eliminated through the use of two frequencies, but both C/A-code, and Y-code receivers can eliminate most of this error using software modelling.[31]

Clock and Ephemeris Error

As shown in Figure C-7, the atomic clocks on board each GPS satellite are designed to provide highly accurate timing specifications. Even a small amount of inaccuracy, however, combined with the fact that the estimated orbital positions, or ephemeris, of each satellite are also not exact, can cause a certain amount of error in a receiver's position solution.

Multipath Errors

Multipath errors occur when incoming GPS signals bounce off a reflective surface such as a building or a body of water before reaching a user's receiver. For highly specialized receivers that are able to eliminate other error sources, pseudorange and/or carrier-phase multipath is frequently a dominant error source.

Receiver Errors

GPS receivers themselves introduce several sources of error to the measurement of satellite ranges. Thermal noise produced by the environment and the various components within a receiver cause small random errors. Received signal to noise ratio, quantization of the analog to digital converter, and the type of tracking loop used by a receiver are also determining factors in the noise level. Typical receiver errors can be as little as 1 centimeter or as large as several meters. This error is quite random in nature and is often reduced by averaging or smoothing over a short period of time.

---

[30] *Space Vehicle Nav System and NTS PRN Navigation Assembly/User System Segment and Monitor Station,* Interface Control Document MH08-00002-400, Revision F, 25 July 1977.

[31] For a typical C/A-code receiver, the remaining tropospheric ranging error amounts to approximately 0.7 meters (1σ). Higher quality C/A-code receivers, and Y-code receivers eliminate all but 0.2 meters of this error.

Dilution of Precision

Dilution of precision, or DOP, is a term that describes the effect of satellite geometry on positioning, timing, and velocity accuracy. Any positioning system that relies on pseudoranging will be affected by the angular spacing between the known points that are used to measure from. The GPS constellation has been designed to give users at least four satellites in view with good geometric spacing, but terrain and man-made structures can occasionally block a receiver's view of some satellites, especially those near the horizon, making the dilution of precision less than ideal.

IMPROVING THE CAPABILITIES OF GPS

Even before the implementation of SA in 1990, many potential GPS users envisioned a need to improve the accuracy of the system, as well as some of its other specified characteristics. Although GPS accuracy has just been discussed, other characteristics such as integrity, availability, continuity of service, and resistance to radio frequency (RF) interference require further elaboration.

Integrity

Integrity, as defined by the *Federal Radionavigation Plan*, is the ability of a navigation system to provide timely warnings to users if and when the system should not be used. The integrity function of a navigation system involves monitoring the system's errors and, if specified protection levels are estimated to be exceeded, giving a warning to the user that the system cannot be used for navigation. In the case of GPS, integrity is maintained by monitoring the signal emanating from each satellite and determining if the pseudorange accuracy meets specified performance criteria for a given application.

Two statistical measures of integrity are often used. One measure relates the probability that a hazardously misleading error will occur and the probability that this error will go undetected (1 minus P_{HE} times P_{MD}, where P_{HE} is the probability of hazardous error and P_{MD} is the probability of missed detection). The second measure of integrity is simply the time a navigation system takes to warn the user that a hazardous error exists (time-to-alarm) There is currently no specified integrity value for either the GPS SPS or the PPS.

Availability

The availability of a navigation system, which is also defined in the *Federal Radionavigation Plan*, is the percentage of time that the services of the system are useable. Availability is an indication of the ability of a system to provide useable service within the specified coverage area. For GPS, "useable service within the specified coverage area" means that at least four satellites must be visible to a user's receiver anywhere on or near the Earth, and the satellites must be providing the required positioning accuracy for the user's application. Some GPS applications, such as static surveying, do not require continuous availability. Others, such as air navigation, can require that GPS signals be available 99.999 percent of the time. The average availability of four or more GPS satellites in view of a given receiver, at SPS accuracy levels, is currently specified as 99.85 percent.[32]

Continuity of Service

Continuity of service, which also is referred to as reliability, is the ability of a navigation system to provide required service over a specified period of time without interruption. The level of continuity is expressed in terms of the probability of not losing the radiated guidance signals.[33] Where warranted, continuity of service is achieved by using redundant transmitters and monitors. Continuity of service and availability go together in that availability is the probability that a system will be in service when it needs to be used, and reliability is the probability that the system will continue to provide service. The global average reliability for GPS is specified as 99.97 percent.[34]

Resistance to RF Interference

The accuracy of a GPS receiver can be degraded in the presence of unwanted interfering signals from terrestrial or other sources. In extreme cases, the receiver is unable to provide any useful navigation or positioning capability. Unwanted and unintentional sources of interference exist, such as the third harmonic of some UHF transmitters, which many civilian users may be unaware of. Military users are also concerned with unintentional interference, but they are more concerned with deliberate efforts to prevent the use of navigation signals through jamming. While no receiver can be made entirely immune to interference (intentional or otherwise), steps can be taken in the design of the receiver to

---

[32] This specified value is the average global availability for a 30-day period, assuming that three satellites have been removed from service on 1 of the 30 days, and assuming a total of 4 satellite down days. Depending on the health of the constellation at any given time, and a users location on the globe, observed SPS-level availability may be better or worse than this average. Source: Assistant Secretary of Defense for Command, Control, Communications, and Intelligence, *Global Positioning System Standard Positioning Service Signal Specification* (Washington, D.C.: U.S. Department of Defense, 8 December 1993), p. B-10.

[33] International Civil Aviation Organization (ICAO), *International Standards and Recommended Practices, Aeronautical Telecommunications*, Annex 10, to the Convention on International Civil Aviation, Volume 1, 22 October 1987.

[34] The full set of assumptions used to determine this value can be found in: Assistant Secretary of Defense for Command, Control, Communications, and Intelligence, *Global Positioning System Standard Positioning Service Signal Specification*, Section 4.0 — Service Reliability Characteristics, pp. B-11 through B-14.

provide some protection against interfering signals. Although quantitative measures of resistance to RF interference, such as jammer-to-signal ratio (J/S) measured in decibels (dB) do exist, these values are very specific to a user's equipment and the signal environment in which it is operating. Therefore, no meaningful specifications for GPS as a complete system can be given.

Augmentations and Enhancements

Many techniques and technical systems designed to improve the capabilities of the basic GPS have been proposed, are under development, or are already in operational use. These techniques range from the differential augmentation of the basic system, to software and hardware enhancements within GPS user equipment, to the integration of GPS user equipment with another navigation/positioning system. Examples of each of these major areas of GPS improvement are discussed below.

Differential GPS

Differential GPS (DGPS) is the most widely used method of GPS augmentation and can significantly improve the accuracy, integrity, and availability of the basic GPS. In fact the term "augmentation" has almost become synonymous with DGPS. DGPS is based upon knowledge of the highly accurate, geodetically surveyed location of a GPS reference station, which observes GPS signals in real time and compares their ranging information to the ranges expected to be observed at its fixed point. The differences between observed ranges and predicted ranges are used to compute corrections to GPS parameters, error sources, and/or resultant positions. These differential corrections are then transmitted to GPS users, who apply the corrections to their received GPS signals or computed position. Figure C-8 further illustrates this concept.

Depending on the user application, DGPS reference stations can be permanent, elaborate installations or small, mobile GPS receivers that can be moved to various well-surveyed locations. The equipment used to broadcast differential corrections, the type of radio datalink used, and the size of the geographic area covered by the DGPS system, also vary greatly with the application. No matter what type of system is used, however, the navigation and positioning capabilities that will be available to any DGPS user within the covered area will be much better than what is available from a stand-alone GPS receiver using either the standard positioning service or the precise positioning service.[35]

---

[35] The term "stand-alone" refers to a receiver that determines position from only the SPS or PPS signal without any augmentation.

Differential GPS Operation

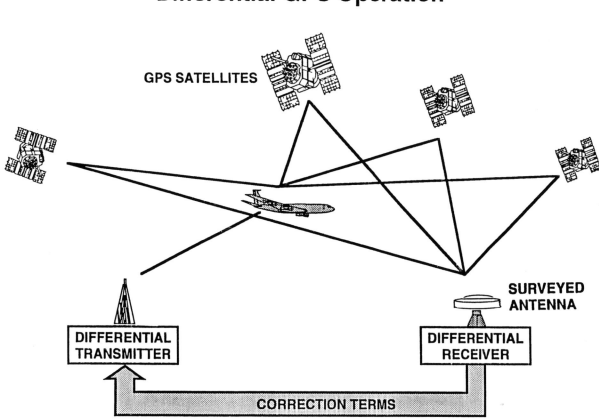

Figure C-8 Differential GPS concept. (Courtesy of the Aerospace Corporation)

Carrier Phase (Interferometric) GPS

In addition to the use of C/A-code, Y-code, or both as measurements of pseudorange for obtaining a position solution, many GPS receivers also measure the L-band carrier phase itself. This enhancement technique can produce very high precision measurements, sometimes as good as 1 to 5 millimeters and, thus, is valuable for high- performance applications. The carrier phase data is used almost exclusively in an interferometric mode, where the phase data from two receivers are processed together to solve for the baseline between them. This eliminates atmospheric errors, and when combined with DGPS, can result in sub-centimeter positioning accuracies.

The difficulty with using carrier phase tracking is the necessity to solve for an unknown quantity termed the integer or cycle ambiguity. Reliable techniques for using carrier phase data in static surveying applications have existed, however, since the mid 1980s. More recently, ambiguity resolution techniques adapted to dynamic applications such as aircraft and ship navigation have also been developed. The success of these new algorithms hinges on the ambiguity resolution technique. One very effective technique, known as wide-laning, relies on carrier phase measurements from both the L_1 and L_2 frequencies.[36]

Multi-channel GPS receivers have recently been developed that take advantage of L_1 and L_2 wide-laning to resolve carrier phase cycle ambiguity by squaring the L2 signal or cross correlating L1 and L2 within a single receiver. The term "codeless" has been associated with these receivers because, as with earlier carrier phase techniques using two receivers, knowledge of the Y-code itself is not required.[37]

Pseudolites

A "pseudolite" or pseudo-satellite is a land-based GPS transmitter capable of generating a signal similar to that of an actual GPS satellite. This signal can be received by a user's GPS receiver without the need for additional frequency reception capability. Pseudolites can improve accuracy, integrity, availability, and continuity of service by simply increasing the number of satellite signals available to the receiver. Adding a differential correction to the broadcast signal makes pseudolites even more effective. Like GPS satellites, however, a pseudolite is only effective if it is within the line of sight of a GPS receiver. The signal power of a pseudolite must also be carefully adjusted to avoid interfering with actual GPS signals.

Receiver Autonomous Integrity Monitoring (RAIM)

Receiver Autonomous Integrity Monitoring (RAIM), as the name implies, is a method to enhance the integrity of a GPS receiver without requiring any external

---

[36] Wide-lane ambiguity resolution (wide-laning) is a processing technique developed by civilian DGPS users to process carrier phase data after using codeless techniques to track or "acquire" the carrier phase. With wide-laning, the two carrier frequencies, which are obtained through codeless techniques, are mixed to provide a difference frequency of longer wavelength. Using L_2 and L_1, the wavelength of the difference frequency is about 4.5 times that of L_1, improving the speed and reliability of cycle ambiguity resolution. The wide-laning technique is available to cross-correlation types of receivers today, but at a serious loss in effective carrier-to-noise ratio as compared to a true dual-frequency code tracking receiver, such as a military PPS receiver using the Y-code on both L_1 and L_2.

[37] For more information on the operation of "codeless" receivers, and GPS receivers in general, see: A. J. Van Dierendonck, "Understanding GPS Receiver Terminology: A Tutorial," *GPS World*, January 1995, pp. 34-44.

augmentations. RAIM algorithms rely on redundant GPS satellite measurements as a means of detecting unreliable satellites or position solutions. All RAIM approaches look for inconsistencies in either the raw measurements or in the position solutions derived from these measurements. RAIM techniques are generally most effective when six or more satellites are in view of the receiver. This means that RAIM alone is not always the best way to improve GPS integrity, and other solutions are often required.[38]

Combined Use of GPS and GLONASS

GLONASS is often discussed as a potential means of augmenting the basic capabilities of GPS by providing additional ranging signals to a user, and integrated GPS/GLONASS receivers are available from a limited number of suppliers. GLONASS, or GLObal Navigation Satellite System, which is operated and managed by the military of the former Soviet Union, consists of three segments just as GPS does. The GLONASS space segment also is designed to consist of 24 satellites, but these satellites are to be arranged in three 64.8° orbital planes 19,100 kilometers (11,870 miles) above the Earth, rather than six planes. The full GLONASS constellation is currently scheduled to be completed in 1995.[39]

GLONASS differs most from GPS in the way that the user segment differentiates one satellite from another. Instead of each satellite transmitting a unique PRN code as GPS satellites do, GLONASS satellites all transmit the same PRN code on different channels or frequencies.[40] All of these frequencies, however, are in the L-band spectrum near either the GPS L_1 or L_2 signal, which simplifies the task of designing integrated receivers. There are still two additional differences between the two systems that must be taken into consideration by combined receiver designers. First, GPS and GLONASS use different time standards for system synchronization. GPS utilizes UTC (Coordinated Universal Time) maintained by the U.S. Naval Observatory ($UTC_{[USNO]}$), whereas GLONASS uses the UTC standard kept in the former Soviet Union ($UTC_{[SU]}$). Discrepancies between these two time scales can reach tens of microseconds, which is significant for systems that keep time with better than 1 microsecond accuracy. Secondly, GPS and GLONASS use different coordinate systems. GLONASS positioning is based on the Soviet Geodetic System (SGS 85), while GPS uses the World Geodetic System (WGS 84) for position determination. Discrepancies between these coordinate systems exist, and must be corrected by combined receivers.

---

[38] More detailed information about RAIM can be found in: R. Grover Brown, "A Basic GPS RAIM Scheme and a Note on the Equivalence of Three Raim Methods," *Navigation: The Journal of the Institute of Navigation* 39, no. 3 (1992).

[39] Further information about the status of the GLONASS program is available from the National Air Intelligence Center, Wright-Patterson Air Force Base, Ohio, which routinely monitors GLONASS developments.

[40] This technique is known as FDMA (Frequency Division Multiple Access).

GPS/Inertial Navigation System (INS) Integration

The present GPS can provide a suitably equipped user with a position, velocity, and time solution whose errors are generally smaller than those of most inertial navigation systems (INS).[41] This performance is achieved in all weather, at any time of day, and under a wide range of signal availability and vehicle dynamics. Nevertheless, the integration of GPS with INS can provide a more robust and possibly more accurate navigation service than is possible with stand-alone sensors. In particular, integration may be the only way to achieve the following:

- Maintain a specified level of navigation performance during outages of GPS satellite reception.

- Reduce the random noise component of errors in the GPS navigation solution.

- Maintain the availability of a GPS solution in the presence of higher vehicle dynamics and radio interference than can be tolerated by GPS alone.

The technical basis for considering GPS/INS integration is the complementary nature of the navigation errors for each system operating in a stand-alone mode. The GPS solution is relatively noisy, but stays within its statistical accuracy boundaries (either CEP or 2 drms boundaries) over time. In contrast, inertial navigation errors are not noisy, but grow in proportion to the duration of a mission and the acceleration experienced by the system. One expects that an integrated navigation solution would perform like an inertial navigator whose errors were bounded by the GPS errors. Additional benefits as noted above are also achievable with more complex integration approaches.[42]

GPS and Loran-C

Loran-C, originally developed by the DOD, is a low frequency (90-110 KHz) radionavigation system that is used by the civil maritime and civil aviation communities. Chains of Loran-C transmitting stations cover the continental U.S. and the coastline of Alaska, as well as the coastlines of many other nations. A Loran-C receiver normally

---

[41] Richard L. Greenspan, *GPS/Inertial Integration Overview*, CSDL-P-3256 (Cambridge, Massachusetts: NATO/Agardograph on Aerospace Navigation Systems, The Charles Stark Draper Laboratory, Inc., March 1993).

[42] For an overview of the benefits of both loosely coupled and tightly coupled GPS/INS integration architectures, see: Richard L. Greenspan, *GPS/Inertial Integration Overview*, pp. 7-10.

determines its position by computing lines of position based on radio pulse transmissions from three stations within a chain.[43]

As with GPS/INS integration, the addition of another navigation system provides redundancy. If GPS signal reception is poor due to a lack of satellites in view or due to signal interference, an integrated system can maintain a specified level of navigation performance using only Loran-C. The system integrity and availability of a GPS/Loran-C system is also improved over GPS alone. A study focused on integrity and availability requirements for aviation non-precision approaches has shown that RAIM performance is significantly improved by the presence of Loran-C signals, and availability improves from 99 percent for a GPS receiver with RAIM and a barometric altimeter to 99.7 percent for a GPS/Loran-C receiver with RAIM.[44]

The integration of Loran-C with DGPS has also been proposed as a potential means of improving both integrity and accuracy. Integrity information and differential corrections could potentially be broadcast on Loran-C signals from existing ground-based transmitter stations to GPS/Loran-C receivers. If this proposal proves to be technically feasible, the entire continental United States and Western Europe could potentially be provided with DGPS capability using Loran-C signals.[45]

PERMANENT DIFFERENTIAL GPS AUGMENTATIONS

It is impossible to estimate the number of temporary DGPS networks in use around the world at any given time because of the ease with which they can be established, utilized, and then removed. GPS users such as surveyors and resource monitors may go through this process several times in one day. It is possible, however, to describe some of the permanent DGPS services that are currently operating or are under development by the U.S. government, state and local governments, foreign governments, and the private sector.

---

[43] More information about Loran-C can be found in the *Federal Radionavigation Plan.*

[44] The availability of Loran-C alone for non-precision approaches is approximately 99.75 percent. Source: James V. Carroll, "Availability Performance Comparisons of Combined Loran-C/GPS and Stand-alone GPS Approach Navigation Systems." in *Proceedings of the IEEE Position Location and Navigation Symposium* (Las Vegas Nevada, April 1994), pp. 77-83.

[45] Lambert J. Beekhuis and Hein J. Anderson, "EuroFix and the Effect of Cross Rate Interference," in *Proceedings of ION-GPS 93: 6th International Technical Meeting of the Satellite Division of the Institute of Navigation* (Salt Lake City, Utah, September 1993), pp. 721-729.

U.S. Government-Supported Differential GPS

There are currently at least a dozen U.S. federal agencies that operate or plan to operate permanent DGPS networks.[46] Three agencies in particular, the FAA (Federal Aviation Administration), the U.S. Coast Guard, and NOAA (National Oceanic and Atmospheric Administration), plan to provide nationwide DGPS services. Each of these three programs is described briefly below.

FAA Wide-Area and Local-Area DGPS Concepts

The FAA plans to improve the accuracy, integrity, and availability of GPS to levels which support flight operations in the National Airspace System from en route navigation through Category I precision approaches by using a wide-area DGPS concept known as the Wide-Area Augmentation System (WAAS).[47] In June 1994, the FAA released an RFP (request-for-proposal) for the WAAS that calls for a ground-based communications network and several geosynchronous satellites to provide nationwide coverage. The ground-based communications network will consist of 24 wide-area reference stations, two wide-area master stations, and two satellite uplink sites. Differential corrections and integrity data derived from the ground-based network, as well as additional ranging data, will be broadcast to users from the geostationary satellites using an "L_1-like" signal with a frequency of 1575.42 MHz.[48] The RFP calls for the WAAS to be in place by the end of 1997.

Local-area DGPS systems are also being considered by the FAA to support landing operations beyond Category I. The airline industry estimates that there are approximately 120 runways in the United States that will require this type of service through the year

---

[46] According to the U.S. General Accounting Office, nine federal agencies either owned and operated, or planned to own and operate permanent differential GPS base stations by fiscal year 1996. They included: the Army Corps of Engineers; the Bureau of Land Management; the Environmental Protection Agency; the Federal Aviation Administration; the Forest Service; the National Oceanic and Atmospheric Administration; the St. Lawrence Seaway Development Corporation; the U.S. Coast Guard; and the U.S. Geological Survey. Source: U.S. General Accounting Office, *Global Positioning Technology: Opportunities for Greater Federal Agency Joint Development and Use*, GAO/RCED-94-280 (Washington, D.C.: U.S. Government Printing Office, September 1994). At least three additional U.S. federal agencies own and operate permanent DGPS reference stations, including the Department of Defense, the National Aeronautics and Space Administration, and the National Science Foundation.

[47] Category I approaches can be flown when the visibility is no less than 0.5 miles (0.8 kilometers), and the ceiling is no lower than 200 feet (61 meters).

[48] Federal Aviation Administration, System Operations and Engineering Branch. *Wide-Area Augmentation System Request For Proposal*, DTFA01-94-R-21474 (Washington, D.C.: U.S. Department of Transportation, 8 June 1994).

2005.[49] Several promising technologies are currently undergoing extensive testing, but an operational system is not expected to be in use until 1999 or beyond.[50]

U.S. Coast Guard DGPS Service

The Coast Guard currently is establishing a DGPS network that will for the first time, meet the extremely accurate navigation requirements of commercial and recreational mariners in our nation's environmentally sensitive harbor and harbor approach areas.[51] When fully operational in 1996, the system is expected to reduce the number of navigation-related grounding and collision incidents by 50 percent over existing navigation methods. A total of 50 reference stations will be installed at sites along the coastal United States, the Great Lakes, Puerto Rico, Alaska, and Hawaii. Each site will use a marine radiobeacon to broadcast differential corrections and integrity information in the RTCM SC-104 message format.[52] The radiobeacon signals can be received by a device about the size of a computer modem with an antenna similar in size to one used by a GPS receiver. By applying the broadcast differential corrections to a GPS position solution in real-time, a user can achieve navigational accuracy as good as 1.5 meters (2 drms) up to 460 kilometers (250 nautical miles) from the reference station.[53]

The Coast Guard hopes to eventually meet the stringent accuracy requirements of inland waterway navigation with their DGPS network as well. In order to achieve this goal, the Coast Guard has entered into a Memorandum of Agreement with the Army Corps of Engineers that will expand DGPS service throughout the navigable waters of the Mississippi River and its tributaries.[54]

NOAA Continuously Operated Reference Stations

The goal of NOAA's Continuously Operated Reference Station (CORS) program is to implement a single, consistent set of federally funded DGPS reference stations that would provide GPS data to all users in a single common format with continuous monitoring of

---

[49] Federal Aviation Administration, *FAA Draft GPS Transition Plan*, 1994. pp. IV-5 and IV-6.

[50] Federal Aviation Administration, *FAA Draft GPS Transition Plan*, pp. II-29.

[51] U.S. Coast Guard, *U.S. Coast Guard GPS Implementation Plan*, June 1994.

[52] The Radio Technical Commission Maritime (RTCM) SC-104 data message is very similar to the GPS navigation message and uses the GPS parity algorithm. Radiobeacons broadcast this message at frequencies between 285 and 325 KHz.

[53] U.S. Coast Guard, *U.S. Coast Guard GPS Implementation Plan*.

[54] U.S. Coast Guard, *U.S. Coast Guard GPS Implementation Plan*.

reference station position. Each reference site would measure coded and codeless L_1 and L_2 data. This data would then be sent to the CORS Central Facility, where it can be stored on computer disc. Users could then access this data electronically within one hour after it has been measured, providing post-processed positioning accuracy of 5 to 10 centimeters. All Coast Guard, Army Corps of Engineers, and FAA reference stations that are part of the DGPS services described above are designed to be CORS-compatible. In addition, a recent technical report to the Secretary of Transportation has recommended that all future federally provided DGPS reference stations should comply with the CORS standard.[55]

State and Local Government DGPS

A number of state and local governments either have established or plan to establish permanent DGPS reference sites. For example, Riverside County, California, has established two continuously operating, permanent DGPS reference stations as part of the Permanent GPS Geodetic Array. This array, whose participants also include federal agencies, state agencies, other local government agencies, and universities, is used primarily for earthquake monitoring and, perhaps, eventually will be used for earthquake prediction. Riverside County engineers and surveyors, however, also use the array for typical day-to-day surveying applications.

Differential Systems Supported by Foreign Governments and International Organizations

Foreign governments and public sector international organizations are actively developing and utilizing differential GPS networks. Several examples designed to support aviation, maritime, and survey/scientific applications are discussed below.

Maritime DGPS Services

Many countries are currently operating, prototyping, or planning maritime DGPS services similar to the U.S. Coast Guard's. The low cost, combined with the absence of any international frequency allocation problems makes these systems practical for all nations. Since most sea coasts and ports have medium-frequency radiobeacons for direction finding, DGPS services can be added quite simply with the purchase and installation of off-the-shelf GPS equipment. The International Association of Lighthouse Authorities (IALA) coordinates the assignment of frequencies and DGPS reference station identifying numbers,

---

[55] U.S. Department of Commerce, National Telecommunications and Information Administration, *A Technical Report to the Secretary of Transportation on a National Approach to Augmented GPS Services*, NTIA Special Publication 94-30, November 1994.

and is compiling information on maritime DGPS broadcasts worldwide. Currently Sweden, Finland, The Netherlands, Denmark, Iceland, and Germany have complete or nearly complete coastal coverage. Several other countries have prototype or demonstration services including Australia, Canada, China, Norway, and Poland. India and South Africa are planning maritime DGPS services.

International Participation in the FAA's WAAS

In order to eventually develop the WAAS into a Global Navigation Satellite System (GNSS) that is useful to aircraft anywhere in the world, the FAA is encouraging other nations to participate in the program at any level they feel comfortable with.[56] Nations involved at the lowest level will simply utilize the GPS-like WAAS signals without any contribution to the system in the form of ground based wide-area reference stations. Participation at a higher level would involve the installation of wide-area reference stations and possibly wide-area master stations within the sovereign territory of a nation. Even higher levels of involvement are possible if a nation is willing to provide a geostationary satellite for the space segment of the system. Several countries have expressed an interest in WAAS participation, including Canada, Australia, New Zealand, and Japan.[57]

Inmarsat

Inmarsat (the International Maritime Satellite Organization), a not-for-profit international organization that provides global mobile satellite services to the maritime, land- mobile, and aviation markets, has firm plans to augment GPS by placing a navigation payload on board its third generation geostationary communications satellites. Plans call for this payload to broadcast GPS and GLONASS integrity information, ranging information, and wide-area differential corrections on a "GPS-like" L_1 signal centered at 1575.42 MHz. These satellites and their navigation payloads may form the nucleus of the WAAS space segment if the winning team of contractors chooses to use them. Future Inmarsat plans include the possible development of a fully civil GNSS based on light satellite (lightsat) navigation payloads placed in intermediate circular orbits and geostationary orbits.[58]

---

[56] Robert Loh, "Worldwide Seamless WAAS Concepts," Viewgraphs presented at the 1995 National Technical Meeting of the Institute of Navigation, Anaheim, California, 18-20 January 1995.

[57] Japan has already established firm plans to develop two geostationary satellites known as the multi-functional transport satellites (MTSAT's) that will augment GPS air navigation in the Asia/Pacific region. These satellites and their ground monitoring network could potentially become part of WAAS.

[58] Jim Nagle, "Waypoints to Radionavigation in the 21st Century," Viewgraphs presented to the National Academy of Public Administration (NAPA) Panel on GPS, 18 November 1994.

The International GPS Service for Geodynamics

The International GPS Service for Geodynamics (IGS) is a network of more than 50 globally distributed GPS tracking sites that has been established by NASA and other organizations from various nations in order to support geodetic and geophysical research activities.[59] Rather than provide real-time differential corrections to users, the tracking sites are used to produce post-processed GPS orbits, or ephemerides, with an accuracy of 10 to 30 centimeters. Orbits are processed at the IGS central bureau at NASA's Jet Propulsion Laboratory in Pasadena, California, and at other sites within the United States and around the globe. These orbits are typically available on the Internet within a few days after they have been processed.[60]

Private Sector DGPS Services

There are a number of private sector enterprises that now offer differential GPS services to the public at various levels of accuracy and at a wide range of prices. These systems use both space-based and land-based datalinks that are encrypted to provide access to only paying customers. Brief summaries of four of these services are provided below.[61]

Racal Survey

Racal Survey of Surrey England (U.K.) has developed a worldwide, space-based differential GPS service known as SkyFix for use in a number of surveying applications. The ground segment of the SkyFix system currently consists of over 25 reference stations around the globe that determine differential corrections that are sent to users via geostationary satellite. The four satellites currently in use are owned and operated by Inmarsat, and provide worldwide coverage except for the polar regions. Users access the differential corrections broadcast in L-band (1530-1545 MHz) using either an Inmarsat terminal or a specialized SkyFix terminal. Racal Survey advertises a positioning accuracy of 3 to 5 meters using this system.

---

[59] Randolph Ware et al., "Optimizing Global Positioning Infrastructure," University NAVSTAR Consortium (UNAVCO), Boulder, Colorado, December 1994.

[60] More information on the IGS can be found in: J. Zumberge et al., "The International GPS Service for Geodynamics — Benefits to Users," *Proceedings of ION-GPS 94: 7th International Technical Meeting of the Satellite Division of the Institute of Navigation* (Salt Lake City, Utah, 20-23 September 1994).

[61] This is by no means an exhaustive list of all the private sector DGPS services currently available.

John E. Chance & Associates, Inc. (A member of the Fugro Group of Companies)

John E. Chance & Associates, Inc, now affiliated with the Dutch Fugro Group, provides DGPS services to North America and much of the rest of the world with a system known as Starfix II.[62] Starfix II systems operate throughout the world by sending differential corrections from each of the reference sites to a central network control center using leased telephone lines, communications satellites, or both. Differential corrections are broadcast to users via L-Band and C-Band geostationary communications satellites and are received by user equipment that consists of a small (3.8 cm high, 7.6 cm diameter) omnidirectional antenna and a signal downconverter (5.0 × 7.6 × 25.4 cm in size).[63] John E. Chance advertises real-time positioning accuracies of 53 centimeters (2 drms).

John E. Chance also provides continuous DGPS coverage to all of the continental United States and most of North America using the OMNISTAR system. The OMNISTAR system is essentially the same as Starfix II, except that differential corrections are broadcast to OMNISTAR users in RTCM SC-104 format, and an ionospheric model that takes the user's location into consideration is utilized in determining the corrections.[64] This approach is a convenient mechanism for providing differential corrections to users with a variety of GPS receivers.

ACCQPOINT

John E. Chance will also provide DGPS correction data via satellite to ACCQPOINT, an FM subcarrier-based DGPS service based on an alliance between Lecia of Torrance, California, and CUE Network based in Irvine, California. ACCQPOINT plans to eventually install receivers for the John E. Chance data at all 500 radio stations that currently are part of CUE's North American paging network. The pseudorange corrections received at the stations will then be broadcast to users within a reception range of 35 to 85 miles (56 to 136 kilometers) using mobile broadcast service (MBS) technology originally developed in Europe. MBS technology allows conventional FM radio broadcasts to carry digital data, such as differential corrections, by modulating the data on an inaudible subcarrier frequency of 57 KHz at approximately 1100 bits per second. The FM subcarrier signal is received by equipment that is only slightly larger than a standard pager and provides users with an advertised accuracy of approximately 1.5 meters.

---

[62] The original Starfix service provided non-GPS positioning accuracy of approximately 5 meters to the Gulf of Mexico and the mid-western United States.

[63] C-band user equipment is larger, but is used only for special applications.

[64] The RTCM SC-104 format also is the standard that has been chosen for the U.S. Coast Guard DGPS network.

Differential Corrections Inc.

Differential Corrections Inc. (DCI), of Cupertino, California, also provides DGPS services to its subscribers using a 57 KHz FM subcarrier. DCI utilizes radio data system (RDS) technology, also originally developed in Europe, that transmits data at 1187.5 bits per second. As of August, 1994, DCI had 46 FM stations operating in their DGPS network, with another 51 stations scheduled to begin service, effectively covering every major population area in the United States. DCI also operates in several foreign countries and hopes to expand their international service.

The differential correction broadcast to DCI users is determined by reference stations located at each of the FM stations within the network. As with ACCUPOINT's service, the user equipment required to receive the FM subcarrier frequency resembles a typical pager. DCI provides its customers with three levels of accuracy at three different prices. The premium service has an advertised accuracy of 1 meter (2 drms), the intermediate service provides 5-meter accuracy, and the basic service gives a customer 10-meter accuracy.

Appendix D

Accuracy Definitions and Mathematical Relationships

Expressions of accuracy stated in this report, unless otherwise noted, are designated as 2 drms.[1] When referring to horizontal positioning, 2 drms is defined as

$$2\sqrt{\sigma_N{}^2 + \sigma_E{}^2}$$

where $\sigma_N{}^2$ and $\sigma_E{}^2$, are the variances of the north and east position estimates respectively. The quantity

$$\sqrt{\sigma_N{}^2 + \sigma_E{}^2}$$

is generally considered to be the uncertainty in the estimation of the two-dimensional (horizontal) position and is called the distance root mean square positional error. Under the simplifying assumption that $\sigma_N{}^2 = \sigma_E{}^2$ and that the errors are independent and normally distributed, the probability that the positional errors are less than 2 drms is 98 percent. In other words, 98 percent of the time in repeated determinations of the horizontal position, the errors will be less than the 2 drms value. In actuality, the percentage of horizontal positions contained within the 2 drms value varies between approximately 95.5 percent and 98.2 percent, depending on the degree of ellipticity of the error distribution.

Circular Error Probable (CEP) is another common measure of horizontal positioning error. CEP is defined to be **CEP = 0.589 (σ_N + σ_E).** The probability of the actual horizontal position lying inside (or outside) a circle with radius CEP is 50 percent. If it is assumed that positioning errors have a circular normal distribution, then the values of CEP and 2 drms are related as follows: **2 drms = 2.4 CEP.**

Similarly the Spherical Error Probable (SEP) is defined to be **SEP = 0.513 (σ_N + σ_E + σ_h).** The probability of the actual position in space lying inside (or outside) a sphere of radius SEP is 50 percent (σ_h is the square root of the variance in the height).

There are other expressions that are commonly used to quantify the uncertainty associated with determination of position. Three such quantities are PDOP, VDOP, and

---

[1] Ronald Braff and Curtis Shively, "GPS Integrity Channel," in vol. III of *Global Positioning System — Papers Published in Navigation* (Washington, D.C.: The Institute of Navigation, 1986), pp. 258–274.

HDOP (Position, Vertical, and Horizontal Dilution of Precision, respectively). Mathematically

$$HDOP = \sqrt{\frac{\sigma_N{}^2 + \sigma_E{}^2}{\sigma_r{}^2}}$$

where $\sigma_r{}^2$ is the variance of a single (pseudorange) observation. Here the level of significance associated with the recovery of position is tied to the uncertainty of the measurements used. This is a function of geometry between the receiver and the tracked satellites.

Appendix E

Report From Mr. Michael Dyment, Booz•Allen & Hamilton

Final Report to
National Academy of Sciences
Committee on the Future of the Global Positioning System

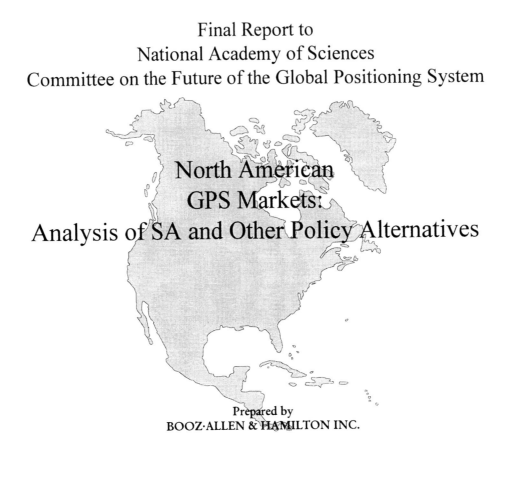

North American
GPS Markets:
Analysis of SA and Other Policy Alternatives

Prepared by
BOOZ·ALLEN & HAMILTON INC.

May 1, 1995

BOOZ·ALLEN & HAMILTON

North American GPS Markets

Since our original projections were developed in September 1992, a number of factors have altered the market for GPS and DGPS products and services. These factors, comprising technology, market and policy elements, are having an influence upon the size and veracity of the GPS and DGPS markets.

*Forces Affecting Markets
Over Ten Year Forecast Period 1994 - 2003*

Technology	Market
• Rapid cost reduction in GPS equipment, brought about by intense competition and accelerated R&D investment. (GPS/DGPS)	• Better performance creating higher user utility, hence accuracy addiction. (GPS/DGPS)
• Advances in LADGPS, WADGPS techniques ... proof of concept, heightened visibility of programs. (DGPS)	• High profile pursuit by market leaders of better performance (more channels in a GPS receiver, better SP characteristics). (DGPS)
• Access to more RF spectrum at better prices -- new digital technologies, better use of available spectrum. (DGPS)	• Maturing markets for in-vehicle navigators in Japan. (GPS)
• Onslaught of integrated products (GPS, CD-ROM, DR), most foreign. (GPS/DGPS)	• Growing consciousness by consumers of benefits of positioning devices -- leading to greater consumer utility. (GPS)

BOOZ·ALLEN & HAMILTON

North American GPS Markets

Since the early 1990's, institutional issues have had a profound impact upon GPS and DGPS market acceptance in North America....an impact that is only now beginning to drive international markets.

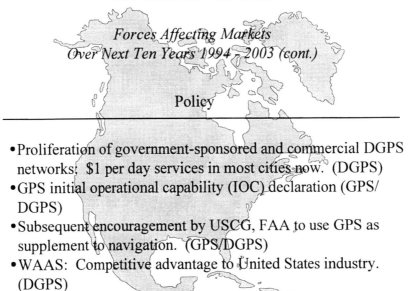

*Forces Affecting Markets
Over Next Ten Years 1994 - 2003 (cont.)*

Policy

- Proliferation of government-sponsored and commercial DGPS networks: $1 per day services in most cities now. (DGPS)
- GPS initial operational capability (IOC) declaration (GPS/DGPS)
- Subsequent encouragement by USCG, FAA to use GPS as supplement to navigation. (GPS/DGPS)
- WAAS: Competitive advantage to United States industry. (DGPS)
- NAFTA approval, international trade sponsored by GATT/WTO and rapid multi-modal transportation infrastructure development (ITS, ports, rail). (GPS/DGPS)
- Increased legislative pressures to mandate DGPS use (i.e. FCC, One Call, CCC, VTS). (GPS/DGPS)

BOOZ•ALLEN & HAMILTON

North American GPS Markets

The general forecast assumptions below were used in our econometric models to quantify market size and composition, and general economic activity in the North American GPS industry (and apply equally with SA On and SA to Zero)

- Economic Assumptions
  - Forecast period is from CY 1994 to 2003 inclusive
  - GNP of the NAFTA countries increases 4% PA in 1994, and 2.25% PA for 1995 to 2003
  - Inflation is held steady at 3% PA over the forecast period in the US and Canada, but 8% in Mexico
- Trade, Import and Export Assumptions
  - Forecasts assume that NAFTA countries will export technology and some end user products, which will be 100% offset by value added imports
  - Foreign technology licenses and resulting royalties are calculated in revenue forecasts
- Revenue Assumptions
  - Industry revenue forecasts are conducted at end of GPS technology industry value added chain - therefore include dealer markups
  - Industry revenue forecasts do not include the increased industrial activity of users resulting from the application of GPS technology and/or products/services, except in the Geomatics sectors.
  - Forecasts include military GPS and DGPS equipment sales and exports
  - Prices of most GPS services, including the provision of DGPS broadcast services, decline at rates of 10% PA over the forecast period
  - Prices of most GPS and communications hardware decline at rates of 5% to 25 % PA, using available benchmark industry data

- Market Assumptions - General
  - End user market data has been derived from sources deemed credible, and include government census data, industry trade association data and at times other market studies
  - Utility and penetration of GPS in many market segments, particularly those which will be driven by product features and prices not yet found in the market, are highly speculative
  - Market adoption as a percent of market potential declines for SPS users over the forecast period with SA on, Market adoption as a percent of market potential increases for DGPS category users over the forecast period with SPS on and off.
- Differential GPS Networks
  - Declared NAFTA country government programs driving DGPS network deployment will continue as committed in 1994, including the US FAA WAAS, US Coast Guard LADGPS, Canadian Gov't CACS, etc.
  - While SPS services are available over the forecast period, DGPS category services are deployed as follows: DGPS-4 reaches 60 % market coverage in 1996; DGPS-3/2 reaches 60% market coverage in 1998; DGPS-1/0 reaches 60% market coverage in 1999
  - Urban DGPS users will have a propensity to use a combination of FM subcarrier, cellular and PCS; Rural and remote DGPS users will have a propensity to use a combination of FM subcarrier and satellite communication
  - The number of commercial and government networks will grow until wide area techniques are found in widespread use, after 1998

BOOZ·ALLEN & HAMILTON

North American GPS Markets

Definitions: Advances in GPS technologies and DGPS techniques have been demonstrated to deliver improved performance. The prospect of SA being turned off in the near future may also be greater. This will improve all accuracy categories.

NAME OF SERVICE	FORECASTED ACCURACY (m 2dRMS) SA ON	FORECASTED ACCURACY (m2dRMS) SA OFF	COMM'L MONTHLY FEES	SP METHOD
SPS (to 1996)	'100	100	FREE	C/A ONLY SA ON
SPS (to 1999)	'100	20	FREE	SA TO ZERO
SPS (to 2001)	'100	'15	FREE	DUAL FREQ SPS
SPS (to 2003)	'100	'5	FREE	IMPROVED OCS
DGPS4	'10	'5	FREE	CODE DIFF
DGPS3/2	'1	'.5	$50-150	CODE DIFF
DGPS1/0	'.02	'.01	$250-500	CODE/NETW

BOOZ•ALLEN & HAMILTON

North American GPS Markets

Defined applications of DGPS categories continue to grow as the utility of GPS is exploited by industries at all levels of the value chain seeking distinct competitive advantages.

Accuracy	Surveying/Mapping	Land Vehicle	Marine	Aircraft Fleet	Personal
DGPS1/0	Geod. Control	Earth Moving	Dredging	CAT II/III	Self-Surveys
	Fault Monitoring	Road Grading	Pylon Posit.	Flight Instrum.	
	Legal Surveys	Agriculture			
	Engineering Surv.				
	Geodynamics				
DGPS2	Resrc. Mapping	Facility Surveys	Rig Positioning	Precision Approach	Golfing
	Reconnaissance	Mapping/GIS	Docking	CAT I	Emergency Location
	Utility Mapping	Highway Surveys	Charting		Fishing
	Highway Surveys	AVLS Trains	Buoy Positioning		
	Legal Surveys	Precision Farming	Seismic Surveys		
DGPS3	GIS Data Collection	AVLS Automobiles	Channel Nav.	Initial Approach	Sailing
	One-Call	AVLS Emergency	Cabling	Vehicle Tracking	
	Photo Control	AVLS Public Transport	Research		
	Navigation	AVLS Trucking	Ship Trials		
DGPS4	IVHS Database GIS	Farming	Harbor Entry	En Route	Hiking
		IVHS Navigation	Harbor Approach	Oceanic	
			Area Navigation		

BOOZ·ALLEN & HAMILTON

North American GPS Markets

*Cost of GPS technology of OEM or imbedded applications
will continue to decline, and performance will dramatically
increase in lockstep.*

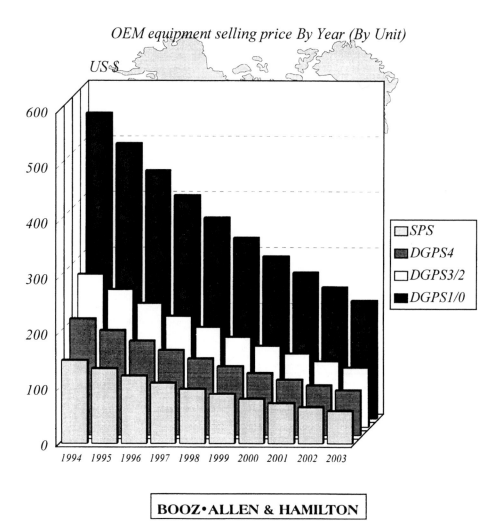

OEM equipment selling price By Year (By Unit)

BOOZ•ALLEN & HAMILTON

North American GPS Markets

The GPS Product area will exhibit similar trends, with avionics and surveying equipment retaining the highest sophistication and selling price.

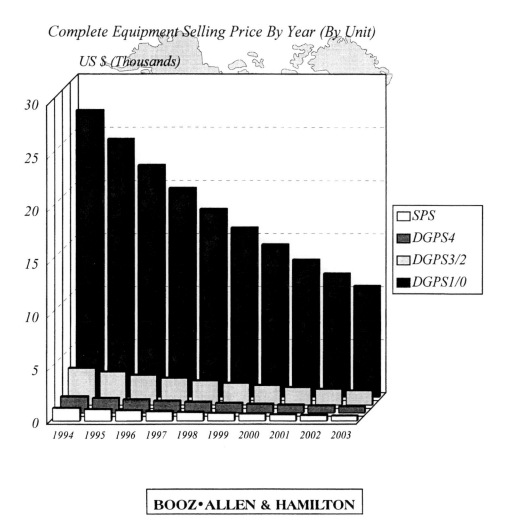

Complete Equipment Selling Price By Year (By Unit)

BOOZ·ALLEN & HAMILTON

North American GPS Markets

The marine market will enjoy complete high end penetration from GPS and DGPS, while the recreational market must wait for higher utility, low cost entries spun out of the land vehicle sector.

Marine Segment Penetration (SA On)
Cumulative 10 Year Market 1994-2003

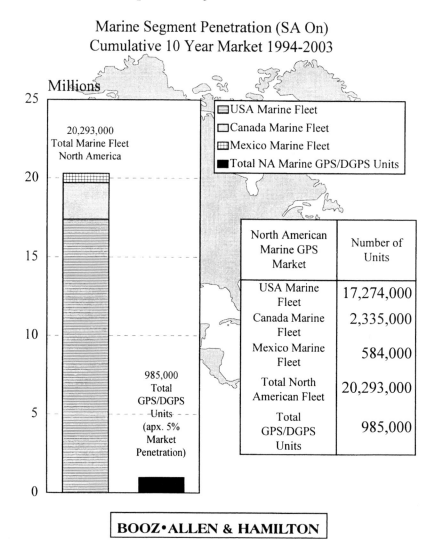

North American Marine GPS Market	Number of Units
USA Marine Fleet	17,274,000
Canada Marine Fleet	2,335,000
Mexico Marine Fleet	584,000
Total North American Fleet	20,293,000
Total GPS/DGPS Units	985,000

BOOZ•ALLEN & HAMILTON

North American GPS Markets

The land vehicle market is comprised of integrated sensors and systems, and CD ROM technologies will set the pace for market adoption in North America.

Land Vehicle Segment Penetration (SA On)
Cumulative 10 Year Market 1994-2003

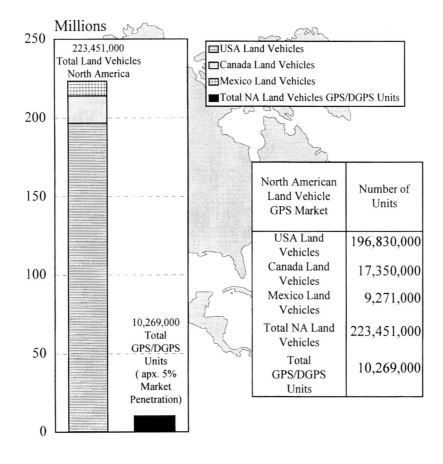

North American Land Vehicle GPS Market	Number of Units
USA Land Vehicles	196,830,000
Canada Land Vehicles	17,350,000
Mexico Land Vehicles	9,271,000
Total NA Land Vehicles	223,451,000
Total GPS/DGPS Units	10,269,000

BOOZ•ALLEN & HAMILTON

North American GPS Markets

The surveying, mapping and GIS markets continue to enjoy the highest penetration percentages of all markets. GIS is only now beginning, with networks such as ACCQPOINT and DCI now available nationally.

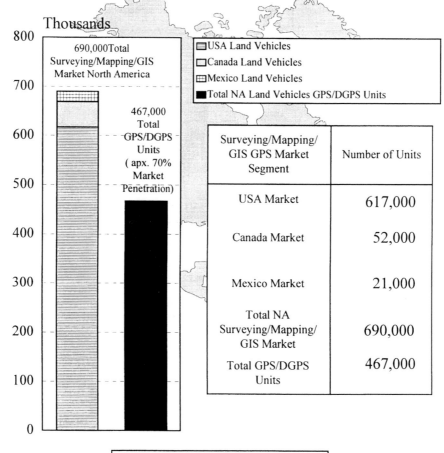

Surveying/ Mapping/GIS Market Segment Penetration (SA On)
Cumulative 10 Year Market 1994-2003

Legend:
- USA Land Vehicles
- Canada Land Vehicles
- Mexico Land Vehicles
- Total NA Land Vehicles GPS/DGPS Units

690,000 Total Surveying/Mapping/GIS Market North America

467,000 Total GPS/DGPS Units (apx. 70% Market Penetration)

Surveying/Mapping/ GIS GPS Market Segment	Number of Units
USA Market	617,000
Canada Market	52,000
Mexico Market	21,000
Total NA Surveying/Mapping/ GIS Market	690,000
Total GPS/DGPS Units	467,000

BOOZ·ALLEN & HAMILTON

North American GPS Markets

The air transport market will remain in turmoil for several years as WAAS and other FAA driven programs set the pace for GPS acceptance.

Aircraft Fleet Market Segment Penetration (SA On)
Cumulative 10 Year Market 1994-2003

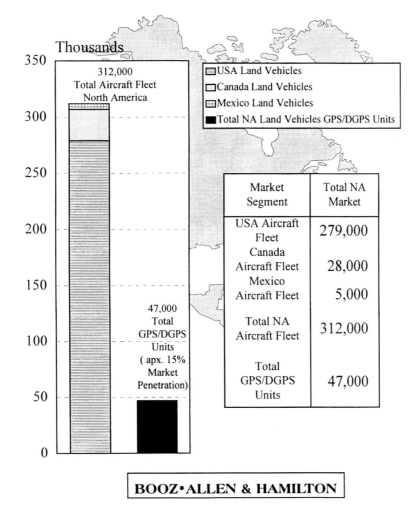

Market Segment	Total NA Market
USA Aircraft Fleet	279,000
Canada Aircraft Fleet	28,000
Mexico Aircraft Fleet	5,000
Total NA Aircraft Fleet	312,000
Total GPS/DGPS Units	47,000

BOOZ•ALLEN & HAMILTON

North American GPS Markets

Careful analysis indicates that with SA On, the largest markets will utilize DGPS corrections, provided by government and private services.

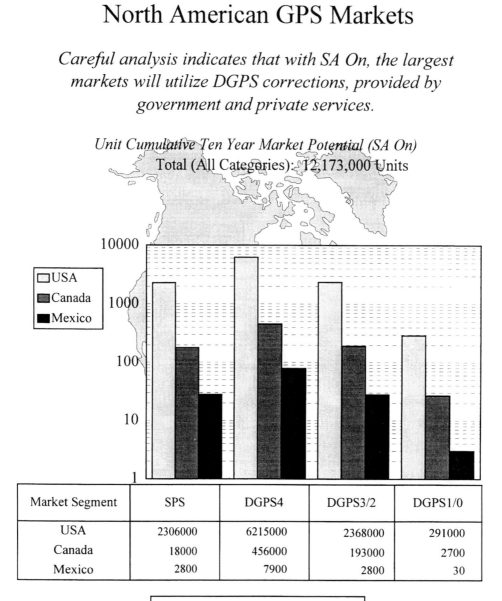

Unit Cumulative Ten Year Market Potential (SA On)
Total (All Categories): 12,173,000 Units

Market Segment	SPS	DGPS4	DGPS3/2	DGPS1/0
USA	2306000	6215000	2368000	291000
Canada	18000	456000	193000	2700
Mexico	2800	7900	2800	30

BOOZ•ALLEN & HAMILTON

North American GPS Markets

While the GPS product market dominates projections in the mid 1990's, the picture changes over time. Soon integrated systems and network services become important components of the overall product mix.

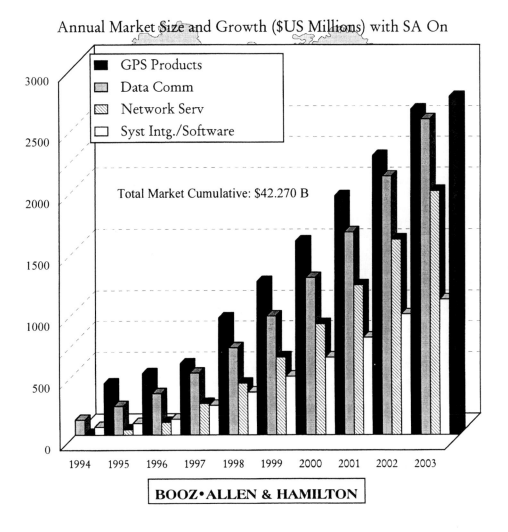

Annual Market Size and Growth ($US Millions) with SA On

- GPS Products
- Data Comm
- Network Serv
- Syst Intg./Software

Total Market Cumulative: $42.270 B

BOOZ•ALLEN & HAMILTON

North American GPS Markets

It is possible that the DoD will relax the present policy on SA, an event that will have a profound impact upon GPS markets in most sectors. Some of the market factors with SA to Zero are indicated below.

- Acceleration in the trend to better accuracies. With higher utility available at throw of switch, most identified markets will mature and new niches will be fostered.
- More rapid international acceptance of DGPS and GPS (civil control would help), generating opportunities for manufacturers.
- Existing GPS products work phenomenally better.
- DGPS networks work better.
- DGPS4 almost disappears as a DGPS performance category.
- WAAS -- CAT I assured.
- R&D emphasis will shift to DGPS1/0.
- Network business restructures around greater accuracies, delivers DGPS1/0 to market 2-3 years sooner.

BOOZ•ALLEN & HAMILTON

North American GPS Markets

Other satellite enhancements could follow in the late 1990's.
Market factors with improved OCS and dual frequency
standard positioning service are indicated below.

- Continued acceleration in the trend to better accuracies.
- Complete elimination of low end DGPS4 service over the forecast period.
- Enhancement of DGPS-3/2/1/0 services through increased accuracy and range of performance (broadcast services still limited by range of real time communications networks).
- DGPS networks work better.
- Wide area techniques will be more assured of better performance with less network sites.
- R&D emphasis will shift to a combination of DGPS1/0 using wide area techniques and from manufacturers, new dual frequency receiver developments.

BOOZ•ALLEN & HAMILTON

North American GPS Markets

With SA to Zero, all categories of performance increase
except DGPS4, which loses to other categories.

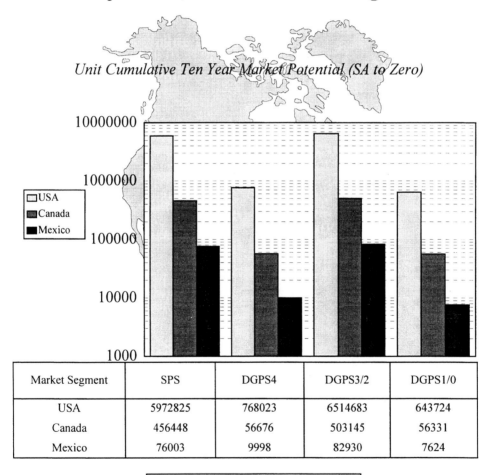

Unit Cumulative Ten Year Market Potential (SA to Zero)

Market Segment	SPS	DGPS4	DGPS3/2	DGPS1/0
USA	5972825	768023	6514683	643724
Canada	456448	56676	503145	56331
Mexico	76003	9998	82930	7624

BOOZ•ALLEN & HAMILTON

North American GPS Markets

All categories of GPS market increase. GPS equipment manufacturing takes a dramatic jump.

Annual Market Size and Growth ($US Millions) with SA to Zero

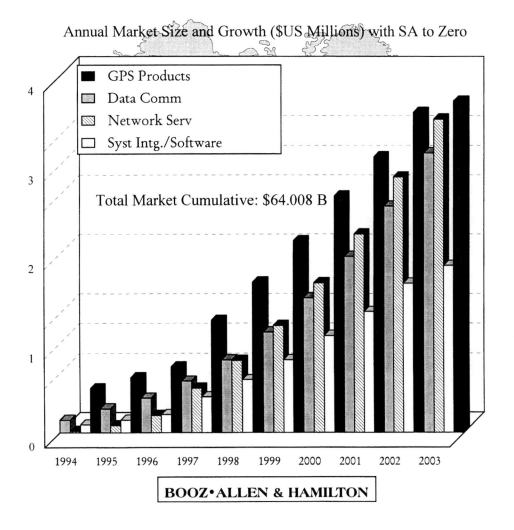

Total Market Cumulative: $64.008 B

Legend:
- GPS Products
- Data Comm
- Network Serv
- Syst Intg./Software

BOOZ•ALLEN & HAMILTON

North American GPS Markets

The impact of policy alternatives upon market composition will be meaningful. Changes in market composition are highlighted below.

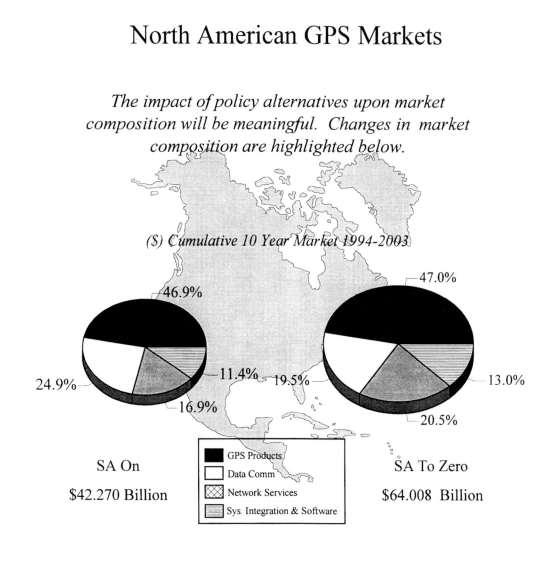

($) Cumulative 10 Year Market 1994-2003

SA On

$42.270 Billion

SA To Zero

$64.008 Billion

Legend:
- GPS Products
- Data Comm
- Network Services
- Sys. Integration & Software

BOOZ•ALLEN & HAMILTON

North American GPS Markets

Comparison of the data by markets shows a variational impact of the SA policy upon each

Impact of Policy Alternatives Upon Market Composition (Units)
Cumulative 10 Year Market 1994-2003

		SPS	DGPS 4	DGPS 3/2	DGPS 1/0
Marine	SA ON	372	334	380	6
	SA TO ZERO	759	50	764	66
Land Vehicle	SA ON	2740	8491	2864	262
	SA TO ZERO	7436	1028	8267	664
Surveying/ Mapping	SA ON	94	143	150	106
	SA TO ZERO	162	15	222	129
Aircraft Fleets	SA ON	29	24	22	17
	SA TO ZERO	56	3	45	27
Personal	SA ON	62	195	66	10
	SA TO ZERO	242	34	271	29

BOOZ•ALLEN & HAMILTON

North American GPS Markets

In summary, the market for GPS and DGPS products and services continues to be an attractive opportunity for manufacturers, developers and systems integrators. With SA to Zero, economic opportunity increases substantially.

- Substantial change in market demographics, led by taste for better accuracies, already a driver in today's policy world.
- SA to Zero: Accelerates market acceptance, opens new markets. Boost to manufacturers of GPS/DGPS products.
- SA to Zero: At least 3 million new users in NAFTA countries by 2003. Growth in all segments of GPS industry.
- SA to Zero: DGPS network services will benefit from higher market momentum and trend toward sub-meter accuracies in many commercial segments.
- SA to Zero: Substantial economic benefits for consumers, commuters and commercial markets. A $10-20 Billion proposition to industry and taxpayers.
- Dual Frequency Commercial: Adds utility, enhances range of operation of DGPS and particularly, WADGPS. Magnifies benefits of "SA to Zero" policy.
- Enhanced GPS satellite service and more realistic policy regarding SA could lead to an additional $20 billion in industrial activity in North America.

BOOZ·ALLEN & HAMILTON

Appendix F

Report From Dr. Young Lee, The MITRE Corporation

MITRE

23 January 1995
F061-L-015

Allison C. Sandlin, PhD
Study Director
National Research Council
Commission on Engineering and
 Technical Systems
2101 Constitution Avenue
Washington, D. C. 20418

Subject: Effect of Eliminating Selective Availability (SA): Impact on Receiver
            Autonomous Integrity Monitoring (RAIM)

Dear Dr. Sandlin:

In response to your request, I have examined the effects of eliminating Selective
Availability (SA) on RAIM. The following summarizes my findings.

Four RAIM augmentations were investigated. They are:

1. Redundant pseudoranges

2. Redundant pseudoranges + altimeter input

3. Redundant pseudoranges + accurate clock

4. Combination of all above

The altimeter input provides another range source. An accurate crystal or small
atomic clock is calibrated when RAIM is available, and used if RAIM becomes
unavailable.

Effects of eliminating SA on RAIM are considered only for the en route and
nonprecision approach phases of flight. The effects are not considered for
precision approach because the required accuracy for that phase of flight is too high
to meet even with elimination of SA.

For all of the above RAIM augmentations, availability and outage durations were
calculated for routes between major city pairs for en route navigation and at
representative terminal areas for nonprecision approach.

Allison, C. Sandlin, PhD
Page 2

23 January 1995
F061-L-015

The results indicate that if selective availability is set to zero, RAIM availability and outage durations will be significantly improved. However, as shown in the Air Navigation Requirements Table in NRC report, the required availability for the FAA's Wide Area Augmentation System is 99.999 percent for GPS to be used as a primary navigation system in the en route and nonprecision approach phases of flight. The results of the above analysis indicate that for 21 operational satellites this level of availability cannot be achieved by RAIM alone even when selective availability is set to zero. Achieving the required availability would require at least 24 satellites operational most of the time and perhaps a dual frequency receiver to correct for ionospheric induced errors.

If you have any questions, please contact Dr. Young C. Lee at (703) 883-7605.

Sincerely,

Dr. Young C. Lee
Lead Engineer
Collision Avoidance, Navigation
and Surveillance Systems

YCL/cp

Enclosure

Effect of Eliminating Selective Availability (SA): Impact on RAIM

Receiver autonomous integrity monitoring (RAIM) is a method for ensuring the integrity of GPS through the use of redundant satellites. For a GPS position solution the pseudoranges of at least four satellites are required. If more than four satellites are in view, the resulting redundancy may be used for integrity by determining the consistency among all of the pseudorange measurements. Thus, in principle, if five satellites are in view, it may be possible *to detect* the presence of a large position error but not to identify which satellite's pseudorange is erroneous. If six or more satellites are in view, it may be possible *to identify* a faulty satellite that is causing a large position error. However, the ability of RAIM to perform the detection and identification functions depends upon the relative geometry between the satellites and the user's location, and upon the nominal pseudorange errors as expressed as a standard deviation. At any place and time the geometry is fixed. Therefore, RAIM improvement would be possible if the standard deviation of pseudorange errors could be significantly decreased. This significant decrease can be obtained by setting selective availability to zero. The effect on RAIM due to setting selective availability to zero can be measured in terms of availability and RAIM outage duration.

RAIM algorithm requirements involve alarm rate, probability of missed detection, and position protection level. False alarm rate must be controlled; otherwise RAIM would be a nuisance. Of course, the missed-detection probability must be low to provide protection when large errors occur. The position protection level means that there will be an extremely low probability that the user's position error will exceed this level without a warning. Table 1 contains a summary of the RAIM algorithm requirements used in the following analysis. These requirements are presently used by the FAA in its evaluation of RAIM.

Four RAIM augmentations were investigated. They are:

1. Redundant pseudoranges

2. Redundant pseudoranges + altimeter input

3. Redundant pseudoranges + accurate clock

4. Combination of all above

The altimeter input provides another range source. An accurate crystal or small atomic clock is calibrated when RAIM is available, and used if RAIM becomes unavailable.

Effects of eliminating Selective Availability (SA) on RAIM are considered only for the en route and nonprecision approach phases of flight. The effects are not considered for precision approach because the required accuracy for that phase of flight is too high to meet even with elimination of SA.

For all of the above RAIM augmentations, availability and outage durations were calculated for routes between major city pairs for en route navigation and at representative terminal areas for nonprecision approach. These are listed in Table 2. Then their average availabilities were tabulated. Separate tabulations were made for

Use of GPS as a supplemental navigation system only requires the former; use of GPS as a primary means of navigation requires both the former and the latter. A supplemental navigation system requires a primary navigation system to be part of the avionics so that in the event of loss of the supplemental system, the pilot can use the primary navigation system. A primary navigation system can operate on its own. Today GPS can be used as a supplemental means of navigation. In the future when GPS is used as a primary means of navigation, RAIM (or some external system) would have to provide the identification function.

Table 3 contains the results for the RAIM detection function when SA is present (pseudorange standard deviation = 33 m) and when SA is absent (pseudorange standard deviation = 4.3 m for dual frequency users and 8.3 m for single frequency users). While the GPS satellite constellation with all 24 satellites operating represents the best case for GPS satellite availability, the probability that all 24 satellites will be operating is estimated to be only about 70 percent. On the hand, DOD guarantees at least 21 satellites to be available with 98 percent probability, and thus the 21 satellite constellation represents a realistic case to address for a primary system.

The results of Table 3 indicate significant improvement when selective availability is set at zero. Since the FAA requires only barometric altimeter input to RAIM for supplemental navigation, the availability improvement from about 90 to 99 percent for supplemental nonprecision approach is very significant when a typical set of 21 satellites are operating.

The results of Table 4 again indicate significant improvement when selective availability is set at zero. The improvement of availability of RAIM identification function for a nonprecision approach is from about 94 to over 99 percent when a typical set of 21 satellites are operating. This is a significant improvement.

The results indicate that if selective availability is set to zero, RAIM availability and outage durations will be significantly improved. As shown in the Air Navigation Requirements Table in NRC report, the required availability for the FAA's Wide Area Augmentation System is 99.999 percent for GPS to be used as a primary navigation system in the en route and nonprecision approach phases of flight. The results of the above analysis indicate that this level of availability cannot be achieved by RAIM alone even when selective availability is set to zero unless perhaps access to dual frequency is available, and the constellation contains at least 24 satellites (Table 4).

Table 1
RAIM Availability Criteria

Function	Requirements
RAIM detection function	1) The presence of a malfunction of a satellite causing the position error protection limit to be violated shall be detected with a minimum probability of 0.999, given that the protection limit is violated and 2) The rate of internal alarms (false or true) shall not be more than 0.002/hr.
RAIM identification function	Upon the occurrence of a malfunctioning satellite with an abnormal range error, RAIM shall be able to detect the occurrence and also correctly identify the satellite before the protection limit is violated with a probability of 0.999.

Table 2
Cases for RAIM Availability Analysis

Phases of flight	User location
En route (Protection limit = 2 nmi) Terminal (Protection limit = 1 nmi)	User on a moving platform: New York to Los Angeles San Fransisco to Narita, Japan Dallas-Fort Worth to Paris
Nonprecision approach (Protection limit = 0.3 nmi)	User at a fixed location: Seattle Chicago Boston Los Angeles Dallas-Fort Worth Miami

Table 3

RAIM Detection Function (5° Mask Angle)

	SA on 21 GPS A	B	C	SA on 24 GPS A	B	C	SA off (single freq user) 21 GPS A	B	C	24 GPS A	B	C	SA off (dual freq user) 21 GPS A	B	C	24 GPS A	B	C
0.3 nmi PL																		
GPS Alone	70.74	33	161	97.86	97	29	94.7	12	44	100	0	0	96.91	7	23	100	0	0
Baro	90.79	21	61	99.84	2	5	99.1	6	15	100	0	0	99.34	5	15	100	0	0
Clock	92.87	19	56	99.83	3	8	99.5	4	14	100	0	0	99.88	2	8	100	0	0
Comb	96.13	21	45	99.86	2	5	99.71	3	14	100	0	0	99.9	2	8	100	0	0
1 nmi PL																		
GPS Alone	93.43	28	49	99.59	5	11	97.54	7	26	99.93	1	3	98	6	26	99.96	1	2
Baro	97.97	9	34	100	0	0	99.52	4	15	100	0	0	99.63	4	14	100	0	0
Clock	98.72	8	17	100	0	0	99.88	2	4	100	0	0	99.98	1	2	100	0	0
Comb	99.06	7	16	100	0	0	99.93	1	4	100	0	0	100	0	0	100	0	0
2 nmi PL																		
GPS Alone	95.8	8	49	99.69	2	6	98.01	6	26	99.96	1	2	98.26	6	26	99.98	0.3	1
Baro	99.59	6	16	100	0	0	99.84	3	10	100	0	0	99.85	3	10	100	0	0
Clock	99.61	4	9	100	0	0	99.98	1	2	100	0	0	99.99	0.3	1	100	0	0
Comb	99.86	1	6	100	0	0	100	0	0	100	0	0	100	0	0	100	0	0

Legend:
A: Availability (%)
B: Average outage duration (min)
C: Maximum outage duration (min)

Table 4

RAIM Identification Function (5° Mask Angle)

		SA on						SA off (single freq user)						SA off (dual freq user)					
		21 GPS			24 GPS			21 GPS			24 GPS			21 GPS			24 GPS		
		A	B	C	A	B	C	A	B	C	A	B	C	A	B	C	A	B	C
0.3 nmi PL	GPS Alone	38.34	52	361	76.19	16	56	67.96	18	86	95.58	6	30	76.92	11	67	97.86	4	12
	Baro	80.89	23	126	98.88	9	17	91.1	14	63	100	0	0	93.12	11	48	100	0	0
	Clock	76.25	26	168	97.88	10	25	92.34	11	67	99.87	2	6	95.54	6	21	99.95	1	3
	Comb	94.3	12	47	99.7	4	9	99.1	6	11	100	0	0	99.53	4	11	100	0	0
1 nmi PL	GPS Alone	66.93	18	215	93.86	8	27	82.28	8	87	98.48	3	13	85.38	6	87	99.01	2	13
	Baro	89.96	14	49	99.57	4	13	94.39	8	40	99.95	1	2	95.19	6	34	99.98	0.3	1
	Clock	91.63	12	49	99.62	3	7	96.37	6	32	99.96	1	2	97.34	4	26	99.98	0.3	1
	Comb	97.42	9	32	100	0	0	99.22	5	0	100	0	0	99.29	5	13	100	0	0
2 nmi PL	GPS Alone	76.36	11	92	97.12	4	17	85.48	6	87	99.01	2	13	87.15	6	87	99.4	2	13
	Baro	93.16	9	40	99.89	2	4	96.34	5	29	99.98	0.3	1	96.8	5	29	100	0	0
	Clock	94.37	9	34	99.85	2	4	97.37	4	26	99.98	0.3	1	97.9	4	26	100	0	0
	Comb	99.17	5	13	100	0	0	99.55	3	0	100	0	0	99.61	3	12	100	0	0

Legend:
A: Availability (%)
B: Average outage duration (min)
C: Maximum outage duration (min)

From: Young Lee, MITRE/CAASD
To: Dave Turner, NRC
Date: Feb. 8, 1995

Dave,

Enclosed is a table showing RAIM availabilities for the additional cases that you wanted to see. I would like to add the following notes to the table.

1. There are one or two GPS stand-alone cases for which availability for 0.3 nmi is higher than that for 1 nmi. You would normally expect that availability would be lower for a smaller protection limit as is the case for alll the other cases in the table. The reason for this apparent discrepancy is that for the case of 0.3 nmi (for nonprecision approach), availability was analyzed at six airport locations in CONUS while availability for 1 and 2 nmi protection limit (for en route) was analyzed for a user flying on three transoceanic routes. There are always slight variations in availability depending on the particular location or route. This apparent discrepancy may disappear if availabilities are examined for a very large number of cases (in terms of location and route) and then averaged.

2. I also find that in many cases, the maximum duration for 1 nmi is larger than that for 0.3 nmi. This is caused by the same reason. In a flight on a moving platform, the user could more easily find not so favorable geometries continuously.

3. As you pointed out, there is one or two GPS/baro cases where availability for 0.3 nmi is higher than that for 1 nmi. As I said before, the reason for it is that barometric altimeter in a nonprecision approach (associated with 0.3 nmi protection limit) gives more accurate altitude data with local barometer setting, and thus can provide higher improvement in availability.

If you have any questions, please feel free to call me at (703)883-7605.

Regards,

                                                                                Young Lee

Table 5

SA off (sigma r = 1.9 m), 5° Mask Angle

RAIM detection

		21 GPS			24 GPS		
		A	B	C	A	B	C
0.3 nmi PL	GPS Alone	98.125	4	11	100.000	0	0
	GPS/Baro	99.549	3	11	100.000	0	0
1 nmi PL	GPS Alone	98.304	6	26	99.978	0.3	1
	GPS/Baro	99.676	4	14	100.000	0	0
2 nmi PL	GPS Alone	98.462	8	26	100.000	0	0
	GPS/Baro	99.854	3	10	100.000	0	0

RAIM Identification

		21 GPS			24 GPS		
		A	B	C	A	B	C
0.3 nmi PL	GPS Alone	83.426	7	67	99.086	2	6
	GPS/Baro	94.363	9	48	100.000	0	0
1 nmi PL	GPS Alone	87.352	7	87	99.407	2	13
	GPS/Baro	95.697	6	34	100.000	0	0
2 nmi PL	GPS Alone	88.138	9	87	99.577	2	13
	GPS/Baro	97.076	5	29	100.000	0	0

Legend:
A: Availability (%)
B: Average outage duration (min)
C: Maximum outage duration (min)

Appendix G

Increased Bandwidth Performance Analysis

To determine more quantitatively the sensitivity of increasing the bandwidth, an analysis was performed using relationships given in the literature for comparing the performance characteristics of the existing C/A-code (narrow band) with that of a wider band signal format.[1]

The code pseudorange error for a narrow correlator design was then fed into a covariance analysis to determine the smoothed pseudorange errors after further carrier-phase smoothing.[2] The following scenarios, which are typical of difficult vehicular applications, were investigated:

- pseudorange accuracy 100 seconds after signal re-acquisition, for zero and high multipath conditions, and

- pseudorange accuracy 10 seconds after signal blockage recovery, for zero multipath and high multipath conditions.

The results are shown in Table G-1. Note that the errors in Table G-1 are pseuodoranges errors (1σ), noise plus multipath.

---

[1] Sources of information: (1) S. N. Karels, T. J. MacDonald, et.al., "Extending Narrow Correlator Space to P(Y) Code Receivers," in *Proceedings of ION GPS-94: 7th International Technical Meeting of the Satellite Divisions of the Institute of Navigation* (Salt Lake City, September 1994). (2) A. J. Van Dierendonck, P. Fenton, and T. Ford, "Theory and Performance of Narrow Correlator Spacing in a GPS Receiver," ION National Technical Meeting (San Diego, January 1992). (3) T. K. Meehan and L. Young, "On Receiver Signal Processing for GPS Multipath Reduction," in *Proceedings Sixth International Geodetic Symposium on Satellite Positioning* (Columbus, Ohio, March 1992), pp. 200-208. (4) L. Weill, "C/A Code Psuedorange: How Good Can It Get?," in *Proceedings of ION GPS-94: 7th International Technical Meeting of the Satellite Divisions of the Institute of Navigation* (Salt Lake City, September 1994); (5) J. W. Sennott, "Multipath Sensitivity and Carrier Slip Tolerance of an Integrated Doppler DGPS Navigation Algorithm," presented at IEEE PLANS-90, March 1990.

[2] J. W. Sennott, "Multipath Sensitivity and Carrier Slip Tolerance of an Integrated Doppler DGPS Navigation Algorithm," presented at the IEEE PLANS-90, March 1990.

Table G-1 Accuracy Recovery Characteristic in Multipath for a Narrow, C/A-Type Code and a Wide-Band, P-Type Signal Format

Signal Type	10 seconds after re-acquisition	100 seconds after re-acquisition
Narrow, C/A-type code High multipath	1.4 meters	1.3 meters
Wide-band, P-type code High multipath	0.4 meters	0.3 meters
Narrow, C/A-type code No multipath	0.37 meters	0.18 meters
Wide-band, P-type code No multipath	0.1 meters	0.057 meters

a. Strong vehicular multipath-to-direct reflection ratio of 0.2, distributed uniformly over full code chip width. Vehicular multipath at code tracking loop output modeled as zero mean Gauss-Markov with a 10-second correlation time.

b. C/A-code receiver with 8-MHz bandwidth and 0.2 chip spacing.

c. Wide-band signal receiver with 20 MHz bandwidth and 1 chip spacing.

d. In all cases 40 dB-Hz carrier-to-noise ratio.

e. In all cases Code loop bandwidth 1 Hz, followed by carrier-smoothed-code filter matched to multipath and ionosphere temporal correlation characteristics.

f. With no multipath, carrier-smoothed code accuracy limited by code minus carrier ionospheric drift.

g. No cycle slips

Under ideal reception conditions, and given sufficient settling time, the pseudorange errors are at the decimeter level for both signal structures. But in the important case of strong multipath, both 10 and 100 seconds after signal blockage, the wide-band, P-type signal is substantially better in performance.

Finally, the relative performance of narrow, C/A-type code and the wide-band, P-type code signal under conditions of in-band interference was examined. In a large number of important civilian applications, a critical requirement is continuous tracking of carrier phase. Beyond the obvious need to recover satellite ephemeris parameters, continuous phase availability allows for smoothing of code pseudorange noise, as well as precise kinematic positioning, Therefore, the susceptibility of phase tracking to in-band interference was of interest. Assuming a phase-tracking threshold of 30 dB-Hz, the tolerable range from a 1-watt, wide-band jammer was computed. The narrow, C/A-type code loss-of-carrier distance was 40 kilometers; the wideband, P-type signal loss-of-carrier distance was 13 kilometers.

Appendix H

Signal Structure Options

Ten signal structure enhancement options were considered by the committee, as shown in Table H-1. Each involves possible changes to L_1 or L_2, as well as a possible signal transmission on a new frequency. The options are listed in priority order.

Table H-1 Signal Structure Options

Option	L₁	L₂	L₄	Advantages Relative to the Current Configuration	Disadvantages Relative to the Current Configuration	Earliest Possible Implementation
1	Y C/A	Y	Pᵃ-like code wide-band signal	Ionospheric correction; improved accuracy; anti-jam; 10-dB improvement over narrow-band in interference rejection; faster cycle ambiguity; fast acquisition; easier direct Y-code acquisition; can track to lower elevation angles than codeless receivers	Must jam two bands; satellite and receiver costs increase; satellite power requirements increase; frequency allocation considerations	IIR
2a	Y C/A	Y	C/A-like code narrow-band signal	Ionospheric correction; improved accuracy; anti-jam; 10-dB improvement over narrow-band in interference rejection; faster cycle ambiguity; fast acquisition; easier direct Y-code acquisition; can track to lower elevation angles than codeless receivers	Must jam two bands; satellite and receiver costs increase; satellite power requirements increase; frequency allocation considerations	IIR
2b	Y C/A		Y with C/A-like code added to null of L₂, narrow-band signal	Ionospheric correction; improved accuracy; anti-jam; 10-dB improvement over narrow-band in interference rejection; faster cycle ambiguity; fast acquisition; easier direct Y-code acquisition; can track to lower elevation angles than codeless receivers	Must jam two bands; satellite and receiver costs increase; satellite power requirements increase; frequency allocation considerations	IIR
3	Y C/A	Y C/A	C/A- or P-like code narrow or wide band signal	Improved accuracy; improved anti-jam for civilians; ionospheric correction; cycle ambiguity	More difficult to deny signal by jamming; more satellite power required	IIF
4	Y C/A	Y C/A	Y-like code (military only) wide-band signal	Improved anti-jam for the military; ionospheric correction for civilians; improved cycle ambiguity; improved direct acquisition of Y-code	Military receiver costs may increase; must jam two bands; may require more satellite power; frequency allocation considerations	IIF

				Baseline	Baseline	Baseline
5	Y / C/A	Y	--			
6	P[a] / C/A	Y	--	Improved accuracy; improved anti-jam; some codeless receivers will have improved performance	current military dual-frequency receivers won't work; some current civilian codeless receivers won't work; must make changes to satellite	IIF
7	Y / C/A	P[a]	--	Improved accuracy; anti-jam; civil ionospheric; correction cycle ambiguity	More difficult to deny signal by jamming; current military dual-frequency receivers won't work; must make changes to satellite	IIF
8	Y / C/A	C/A	--	Civil ionospheric correction; improved cycle ambiguity; some jam resistance	Military receiver costs increase; must jam two bands; satellite power may increase; no dual-frequency military ionospheric correction	(Current) II/IIA
9	P[a] / C/A	P[a]	Y-like code (military only) wide-band signal	Precision; improved anti-jam; provides ionospheric correction for civilian users; improved cycle ambiguity	Military receiver costs increase; must jam two bands; satellite power may increase; possible frequency allocation difficulties; no dual-frequency military-only ionospheric correction	IIF

a. "P" refers to the unencrypted code

OPTIONS 1 AND 2

Options 1 and 2 provide the optimal balance between civilian and military utility. These options were selected by the committee for further study and are discussed in Chapter 3 of this report, along with specific recommendations.

OPTIONS 3 AND 4

Options 3 and 4 include two variants. For both, a C/A-code is added as soon as practical to L_2 transmissions. This would be relatively easy to implement on Block IIR spacecraft. With either option, a new civilian or military signal could be added when practical. In the near term, civilian users would benefit in terms of interference reduction, ionospheric error reduction, and improved reliability of cycle ambiguity wide-laning. With the later enhancement of an additional civilian signal, many of the advantages of Option 1 would be obtained.

However, enabling C/A-code on both L_1 and L_2 raises potential difficulties for military local access denial. Under Option 3, the military would need to jam three separate civilian frequencies, two of which overlap the military frequencies. Both L_1 and L_2 would be affected simultaneously, which could have undesirable consequences for the existing inventory of military receivers.

Under Option 4, a new dedicated military wide-band signal with an encrypted code would be added to provide increased military capability and better segregation of military and civilian services.

OPTION 5

Option 5 is the baseline case. As pointed out earlier in this report, the civilian community currently has many applications where the narrow-bandwidth C/A-code structure is detrimental. Furthermore, the lack of a second frequency with known codes has substantial impact upon precise differential applications as well as on stand-alone applications. Since the Block IIF constellation lifetime could extend into the year 2020 or beyond, it follows that an acceptance of this option could render GPS obsolete.

OPTION 6

Option 6 eliminates encryption on L_1, which allows full civil access to the wide-band P-code, with many potential performance benefits. Anti-spoofing remains on at the L_2 frequency. While enhancing civilian performance, it negatively impacts some existing civilian receivers and most military receivers. Civilian codeless receivers of the cross-correlation variety will need modification to handle processing of P-code and Y-code together. The widely deployed military P-code receiver "Plugger" will loose its anti-spoofing capability, and intentional jamming of L_1 will inhibit two-frequency ionospheric corrections for the military.

OPTION 7

Option 7 provides civilian access to a wide-band signal format, as well as excellent dual-frequency wide-laning and ionospheric corrections. As in Option 6, some changes to military software and hardware will be required to handle the mixed P/Y-code situation on L_1 and L_2. However, this change is compatible with single-frequency military receivers such as the Plugger. Local denial will entail selective jamming and/or C/A-code spoofing on L_1, as well as complete jamming/spoofing on the L_2 band. In a geographic region of denial the military might be without a dual-frequency capability.

OPTION 8

Option 8 emphasizes civilian dual-frequency operation, as well as military A-S operation. Civilians would obtain very good wide-laning capability, but would not get enhanced wide-bandwidth features. Also, the availability of widely spaced frequencies would offer some interference reduction. On the military side, ionospheric correction might be lost in denial-jamming/spoofing situations unless careful cross-aiding from L_1 were employed, and the military would not have a signal solely for their purposes.

OPTION 9

Option 9 essentially gives to civilians the wide-band, dual-frequency capabilities of the military. Clearly, this option would be highly beneficial to the civilian sector, but it would leave most of the military receiver inventory vulnerable to spoofing or even outright loss of navigation capability in denial environments. The most critical military users would have available a new Y-code signal, perhaps of much wider bandwidth and operating on a higher carrier frequency. Such a Y-code signal upgrade is for the late/post Block IIF time period.

Appendix I

Report From Mr. Melvin Barmat, Jansky/Barmat Telecommunications, Inc.

JANSKY/BARMAT TELECOMMUNICATIONS, INC.

1899 L STREET, N.W.

SUITE 1010

WASHINGTON, D.C. 20036

(202) 467-6400

(Fax) (202) 296-6892

Preliminary Assessment of

Frequency Bands for Use by

a New GPS Signal

for the

Committee on the Future of Global Positional System

of the

National Research Council

Melvin Barmat

January 4, 1995

Preliminary Assessment of
Frequency Bands for Use by
a New GPS Signal

ABSTRACT

Because of the congestion in the use of the 1-2 GHz frequency band, an initial investigation was undertaken to evaluate the potential availability of spectrum to support a new GPS signal. The band 960 to 1950 MHz was considered and several sub-bands appear to have promise for the proposed new signal. In-depth investigation and coordination is necessary before a specific band can be selected.

Preliminary Assessment Of
Frequency Bands For Use By A New GPS Signal

Statement of Problem

A new GPS navigation signal has been suggested for use by the civil sector. In light of the intense congestion of the 1-2 GHz frequency band, an assessment was undertaken to identify possible frequency bands that could be used for transmissions of this new GPS signal. The minimum bandwidth required would be that needed for a C/A-code, although if sufficient appropriate spectrum were available, a wider bandwidth signal may be considered.

Major Spectrum Constraints

o   The new GPS signal (L4) would be primarily used to aid in correcting for ionospheric-caused group delay. The advice of a leading expert (J.A. Klobuchar of the Air Force Geophysics Lab) was sought to suggest appropriate frequency separations from the GPS L1 (1575.42 MHz) signal. His memo is Attachment A.

o   The new GPS signal (L4) should not cause the cost of civil GPS receivers to be increased substantially and therefor should preferentially use frequencies between L1 (1575.42 MHz) and L2 (1227.60 MHz). Alternatively, L4 should employ frequencies no more than 300 MHz higher than L1.

o   Neither the new GPS L4 signal nor its out-of-band components shall be the cause of unacceptable interference to systems employing the same or adjacent frequencies. Likewise, the new signal shall not be required to accept harmful interference from other systems employing the same or adjacent frequencies.

o   Because of the above, and since vehicles will often use the new GPS signal, frequency bands currently allocated for transmission of communications[1] from vehicles should be avoided.[2] Bands allocated to the space-to-Earth Mobile Satellite Service should also be avoided since the transmitted power (both in-band and nearby out-of-band) of such satellites will cause harmful interference to GPS signals.

o   The frequency band to be used for the new GPS L4 signal must be either shared on an acceptable interference basis with another service or alternatively, an appropriate part of a band must be cleared[1] for use solely by the new GPS L4 signal. (Note the 1-3 GHz band is in great demand

---

[1]   It may be possible for the new GPS signal to share a frequency band with radiolocation or radionavigation systems employing short pulse - duration transmissions. Further study is needed.

[2]   Interference problems can occur when a communications transmitter and GPS receiver operating at the same or nearby frequencies are on the same or nearby vehicles. An example of this problem has surfaced for aeronautical mobile satellite transmitters interfering with Glonass receivers operating in nearby frequencies on the same aircraft.

for mobile communications services and there is no "free" spectrum at this time.)

o   The new GPS signal will be used worldwide.

Frequency Band Considerations

Attached Table 1 considers the possible use of frequency subbands between 1 and 2 GHz for the new GPS L4 signal. The table includes comments on the advantages and disadvantages of each subband for the new GPS signal. Table 2 is a copy of the relevant portions of the Table of Frequency Allocations of the international Radio Regulations.

Although considerable investigation is needed before a decision is made in selecting a band for L4, Table 1 shows that several bands in the 1-2 GHZ range look promising.

Course of Action

Obviously, prior to selecting a band for the new GPS signal, in-depth investigations regarding existing use of the potential bands and detailed analyses regarding sharing of frequencies (including consideration of out-of-band emissions) need to be undertaken. Discussions with representative users of affected bands are also recommended.

In order to obtain international status for the new GPS frequency band, it will be necessary that the band be allocated to the Radionavigation-Satellite Service through the ITU process.

Frequency allocations are made at international treaty conferences called World Radio Conferences (WRC). In order to help ensure that a WRC acts to allocate a band for L4, it is suggested that at an appropriate time both ICAO and IMO be solicited for support. If a new Radionavigation-Satellite allocation is needed for the GPS-L4 signal, the earliest probable date for accomplishing that goal is believed to be 1999.

TABLE 1

Comments on Potential Use

of Frequency Bands for a

New GPS Signal

TABLE 1

Table of Potential Spectrum for New GPS Carrier

Band (MHz)	Disposition/Comment	Discussion
960-1215	May be Possible. Needs consultation with aeronautical interests.	This 255 MHz bandwidth is allocated on a worldwide basis to Aeronautical Radionavigation services. A wide number of civil and military aeronautical systems employ this band including DME, TACAN, JTIDS, TCAS, Mode-S, IFF, etc. However, the use of some of these systems may become obsolete because of widespread employment of GPS. Thus, bandwidth may become available in the future. Moreover, since a C/A code GPS-L4 would only need a few MHz of bandwidth, it would seem in the self-interest of the aeronautical community to make such a small amount of spectrum available.
1215-1240	Good Possibility. Needs further study, including GPS interaction and interference effects of radar.	Worldwide allocation to Radiolocation and Radionavigation-Satellite services. The GPS-L2 signal is located in this band at 1227.6 MHz. Since the P-code occupies $\pm$10 MHz, the bands 1215-1217.6 MHz or 1237.6-1240 MHz may be available for the GPS-L4 signal. If 1 or 2 MHz of additional spectrum is needed, it may be available in one of the adjacent bands. However, the transmission and reception of two adjacent frequency signals (L2 and L4) often gives rise to intermodulation interference problems. Further study is obviously needed.

In addition, it may be possible to operate a C/A-code signal and a P-code signal co-frequency at 1227.6 MHz (L2), as it is done at 1575.42 MHz (GPS-L1).

In the U.S., the Radiolocation services are for U.S. government long-range air surveillance radars. Worldwide usage needs investigation.

Use of this band has the advantage of not requiring an allocation change via the ITU. |

Band	Possibility	Description
1240–1260	Good Possibility. Needs detailed study including effect of radar on GPS signal.	Worldwide allocations to Radiolocation and Radio-navigation-Satellite. The latter covers the L2 channel for Glonass. However, the Russian gov't has agreed to move Glonass frequencies down from their current band over the next several years. Thus, it is probable that several MHz at the high end of this band will be cleared out and could be used for GPS L4. However, the use of the co-frequency Radiolocation allocation needs further study to verify sharing potential. In the U.S. this band is assigned to DOD for radars and use may be extensive. However, it is encouraging to note that the band has been acceptable for Glonass L2. An advantage of this band for GPS L4 is that no new ITU allocation would be necessary. NASA uses this band on a secondary basis for earth-exploration satellites.
1260–1300	Possible. Needs study, including radar effects.	Worldwide allocation to Radiolocation (radar). In the U.S., this band is assigned to DOD. Secondary amateur satellite service operates Earth-space in this band and could cause localized interference that would be difficult to uncover.
1300–1350	Low Possibility. Needs study, including radar effects.	This band is allocated worldwide to Aeronautical Radionavigation and is used for surveillance radars in air traffic control (ATC) in the U.S. Transponders on the aircraft are used for identification. It is believed this same use is worldwide. In U.S. and elsewhere, ATC surveillance radars may be partially replaced by Mode S systems in the future. Thus, in the long term it may be possible that several MHz of this band could be set aside for GPS L4.

There is a footnote asking administrations to protect the band 1330–1400 MHz for Radioastronomy. Space station emissions are noted to be particularly harmful to these observations.

Band (MHz)	Assessment	Comments
1350-1400	Questionable. would need much study.	The band has a worldwide allocation for Radiolocation. The band is used for long range surveillance radars by the DOD and for air traffic control. There is also a Radio-astronomy footnote, noted above (RR 718), and the fact that passive space research is conducted in 1370-1400 MHz (RR 720) would make this a difficult band for NAS to endorse. The U.S.G. is turning control of 1390-1400 MHz over to non-government use but notes problem of Radioastronomy. Also, the fixed and mobile allocations in Region 1 are potential sources of harmful interference to GPS-L4. Note that GPS L3 at 1381 MHz is in this band but its use is subject to restrictions.
1400-1427	Not Feasible	Primary allocations to Radioastronomy and Space Research (passive) make this band not feasible.
1427-1429 1429-1452	Probably not feasible Needs study.	Mobile and Fixed services are allocated worldwide in these bands. One of the ground rules established in this study is to not attempt to place a GPS receiver on a mobile vehicle where a co-frequency transmitter was possible. However, the U.S.G. will turn over control of 1427-1432 MHz to non-government use in about five years, although some government use will be grandfathered! Note that the adjacent band is allocated on a Primary basis to Radioastronomy.
1452-1492	Not Feasible	The current application of this band to the Broadcasting, Broadcasting-Satellite and Mobile services make this band not feasible for GPS-L4.
1492-1525	Not Feasible	The band is allocated to Mobile services worldwide and to Mobile Satellite (space-Earth) in Region 2. Either allocation would rule out its use for GPS-L4.
1525-1530 1530-1533 1533-1535 1535-1544 1544-1545	Not Feasible	Worldwide allocation to Mobile Satellite (space-Earth). These satellite transmissions would cause unacceptable interference to an L4 GPS receiver in the band.

1545-1555 1555-1559		
1559-1610	Not Feasible	Band used for GPS and Glonass L1 signals. Band edge use would not be acceptable.
1610-1610.6 1610.6-1613.8 1613.8-1626.5 1626.5-1631.5 1631.5-1634.5 1634.5-1645.5 1645.5-1646.5 1646.5-1656.5 1656.5-1660 1660-1660.5	Not Feasible	Worldwide allocation to Mobile Satellite (Earth-space). Uplink transmissions from vehicle to satellite would cause unacceptable interference to GPS receiver on same or near-by vehicle.
1660.5-1668.4 1668.4-1670	Not Feasible	Worldwide allocations to Radioastronomy.
1670-1675	Not Feasible	Worldwide allocations to Metaids, Mobile, etc. would cause unacceptable interference to GPS-L4. In addition, a new air passenger telephone service (ground-to-air) is planned for this band. The U.S.G. is turning over control of 1670-1675 MHz to non-government use and thereby not allowing radiosonde operations in the band. However, the U.S.G. points out the sensitivity of Radioastronomy in the adjacent band.
1675-1690 1690-1700	Not Feasible	Worldwide allocation to Metaids (radiosondes), Metsats (space-Earth) and Mobile services make use of this band not feasible in the near term. Moreover, the band is allocated to mobile satellite (Earth-space) in Region 2 and there are indications the allocation will be extended worldwide.
1700-1710	Not Feasible	Same as above, except for omission of Metaid allocation.

| 1710-1930 | Good possibility. Study needed. |

This 220 MHz band has worldwide allocations for Fixed and Mobile services. However, band usage is in a state of flux. The band 1800-1805 MHz is set aside for air-to-ground link of a new air passenger telephone service. The band 1885-2025 is set aside for a next generation worldwide public land mobile telephone system and the FCC has extended the band down to 1850 MHz for such use in the U.S. The band 1761-1842 MHz is used for a number of USG space services. The U.S. government has announced that it intends to turn over primary use of 1710-1755 MHz to non-government service in 2004, although some U.S.G. use will be grandfathered. 1718.8-1722.2 MHz is allocated to radio astronomy on a secondary basis. The lower end of this band looks promising for GPS L4, although out-of-band emissions of LEO weather satellites need investigation.

TABLE 2

Frequency Allocations in the International Radio Regulations

Table 1 contains comments on the potential utilization of spectrum for the new GPS signal (L4) and is based on the international Radio Regulations. The relevant portion of the Table of Frequency Allocations of the Radio Regulations is attached hereto. Before describing this Table, it is important to note that the international Radio Regulations are treaty obligations (having the force of law) of a country when ratified by the appropriate body in that nation.

For those not familiar with the Radio Regulations, a few explanations regarding Table 2 are necessary. Frequency bands (leftmost bold numbers) are allocated to services by region. Region 1 is Europe (plus Siberia) and Africa; Region 2, the Americas; Region 3 is Asia and Oceana. Services in capital letters (e.g., RADIOLOCATION) have Primary status, while services spelled in lower case letters (e.g., Amateur) have Secondary status. The other numbers in Table 2 associated with an allocation relate to footnotes that cover a range of subjects, although most of them are country-specific exceptions to a Regional allocation. A sample page of footnotes is in Table 3.

MHz
890 – 1 240

Allocation to Services		
Region 1	Region 2	Region 3

890 – 942
FIXED
MOBILE except
  aeronautical mobile
BROADCASTING 703
Radiolocation
	890 – 902
	FIXED
	MOBILE except
	  aeronautical mobile
	Radiolocation
	700A 704A 705
		890 – 942
		FIXED
		MOBILE
		BROADCASTING
		Radiolocation

	902 – 928 FIXED Amateur Mobile except aeronautical mobile Radiolocation 705 707 707A	
	928 – 942 FIXED MOBILE except aeronautical mobile Radiolocation	
704	705	706
942 – 960 FIXED MOBILE except aeronautical mobile BROADCASTING 703 704	**942 – 960** FIXED MOBILE	**942 – 960** FIXED MOBILE BROADCASTING 701
960 – 1 215	AERONAUTICAL RADIONAVIGATION 709	
1 215 – 1 240	RADIOLOCATION RADIONAVIGATION-SATELLITE (space-to-Earth) 710 711 712 712A 713	

MHz
1 240 – 1 452

Allocation to Services		
Region 1	Region 2	Region 3
1 240 – 1 260	RADIOLOCATION RADIONAVIGATION-SATELLITE (space-to-Earth) 710 Amateur 711 712 712A 713 714	
1 260 – 1 300	RADIOLOCATION Amateur 664 711 712 712A 713 714	
1 300 – 1 350	AERONAUTICAL RADIONAVIGATION 717 Radiolocation 715 716 718	

1 350 – 1 400
FIXED
MOBILE
RADIOLOCATION
 	
718   719   720
	1 350 – 1 400
	          RADIOLOCATION
	
	          714 718 720	

1 400 – 1 427	EARTH EXPLORATION-SATELLITE (passive) RADIO ASTRONOMY SPACE RESEARCH (passive) 721 722	
1 427 – 1 429	SPACE OPERATION (Earth-to-space) FIXED MOBILE except aeronautical mobile 722	

1 429 – 1 452
FIXED
MOBILE except
  aeronautical mobile
 	
722   723B
	1 429 – 1 452
	          FIXED
	          MOBILE 723
	
	          722	

MHz
1 452 – 1 530

Allocation to Services		
Region 1	Region 2	Region 3
1 452 – 1 492 FIXED MOBILE except aeronautical mobile BROADCASTING 722A 722B BROADCASTING- SATELLITE 722A 722B 722 723B	**1 452 – 1 492** FIXED MOBILE 723 BROADCASTING 722A 722B BROADCASTING-SATELLITE 722A 722B 722 722C	
1 492 – 1 525 FIXED MOBILE except aeronautical mobile 722 723B	**1 492 – 1 525** FIXED MOBILE 723 MOBILE-SATELLITE (space-to-Earth) 722 722C 723C	**1 492 – 1 525** FIXED MOBILE 723 722
1 525 – 1 530 SPACE OPERATION (space-to-Earth) FIXED MARITIME MOBILE-SATELLITE (space-to-Earth) Earth Exploration-Satellite Land Mobile-Satellite (space-to-Earth) 726B Mobile except aeronautical mobile 724 722 723B 725 726A 726D	**1 525 – 1 530** SPACE OPERATION (space-to-Earth) MOBILE-SATELLITE (space-to-Earth) Earth Exploration-Satellite Fixed Mobile 723 722 723A 726A 726D	**1 525 – 1 530** SPACE OPERATION (space-to-Earth) FIXED MOBILE-SATELLITE (space-to-Earth) Earth Exploration-Satellite Mobile 723 724 722 726A 726D

MHz
1 530 – 1 545

Allocation to Services		
Region 1	Region 2	Region 3

1 530 – 1 533
 	
SPACE OPERATION
  (space-to-Earth)
MARITIME MOBILE-
  SATELLITE
  (space-to-Earth)
LAND MOBILE-
  SATELLITE
  (space-to-Earth)
Earth Exploration-Satellite
Fixed
Mobile except
  aeronautical mobile
 	
722   723B   726A   726D
	1 530 – 1 533
	
	SPACE OPERATION (space-to-Earth)
	MARITIME MOBILE-SATELLITE
	  (space-to-Earth)
	LAND MOBILE-SATELLITE
	  (space-to-Earth)
	Earth Exploration-Satellite
	Fixed
	Mobile 723
	
	722 726A 726C 726D	

1 533 – 1 535
 	
SPACE OPERATION
  (space-to-Earth)
MARITIME MOBILE-
  SATELLITE
  (space-to-Earth)
Earth Exploration-Satellite
Fixed
Mobile except
  aeronautical mobile
Land Mobile-Satellite
  (space-to-Earth)   726B
 	
722   723B   726A   726D
	1 533 – 1 535
	
	SPACE OPERATION (space-to-Earth)
	MARITIME MOBILE-SATELLITE
	  (space-to-Earth)
	Earth Exploration-Satellite
	Fixed
	Mobile 723
	Land Mobile-Satellite (space-to-Earth) 726B
	
	722 726A 726C 726D	

1 535 – 1 544	MARITIME MOBILE-SATELLITE (space-to-Earth) Land Mobile-Satellite (space-to-Earth) 726B 722 726A 726C 726D 727	
1 544 – 1 545	MOBILE-SATELLITE (space-to-Earth) 722 726D 727 727A	

MHz
1 545 – 1 613.8

Allocation to Services		
Region 1	Region 2	Region 3
1 545 – 1 555	AERONAUTICAL MOBILE-SATELLITE (R) (space-to-Earth)	
	722 726A 726D 727 729 729A 730	
1 555 – 1 559	LAND MOBILE-SATELLITE (space-to-Earth)	
	722 726A 726D 727 730 730A 730B 730C	
1 559 – 1 610	AERONAUTICAL RADIONAVIGATION RADIONAVIGATION-SATELLITE (space-to-Earth)	
	722 727 730 731	
1 610 – 1 610.6 MOBILE-SATELLITE (Earth-to-space) AERONAUTICAL RADIONAVIGATION 722 727 730 731 731E 732 733 733A 733B 733E 733F	1 610 – 1 610.6 MOBILE-SATELLITE (Earth-to-space) AERONAUTICAL RADIONAVIGATION RADIODETERMINATION-SATELLITE (Earth-to-space) 722 731E 732 733 733A 733C 733D 733E	1 610 – 1 610.6 MOBILE-SATELLITE (Earth-to-space) AERONAUTICAL RADIONAVIGATION Radiodetermination-Satellite (Earth-to-space) 722 727 730 731E 732 733 733A 733B 733E
1 610.6 – 1 613.8 MOBILE-SATELLITE (Earth-to-space) RADIO ASTRONOMY AERONAUTICAL RADIONAVIGATION 722 727 730 731 731E 732 733 733A 733B 733E 733F 734	1 610.6 – 1 613.8 MOBILE-SATELLITE (Earth-to-space) RADIO ASTRONOMY AERONAUTICAL RADIONAVIGATION RADIODETERMINATION-SATELLITE (Earth-to-space) 722 731E 732 733 733A 733C 733D 733E 734	1 610.6 – 1 613.8 MOBILE-SATELLITE (Earth-to-space) RADIO ASTRONOMY AERONAUTICAL RADIONAVIGATION Radiodetermination-Satellite (Earth-to-space) 722 727 730 731E 732 733 733A 733B 733E 734

MHz
1 613.8 – 1 656.5

Allocation to Services		
Region 1	Region 2	Region 3
1 613.8 – 1 626.5 MOBILE-SATELLITE (Earth-to-space) AERONAUTICAL RADIONAVIGATION Mobile-Satellite (space-to-Earth) 722 727 730 731 731E 731F 732 733 733A 733B 733E 733F	**1 613.8 – 1 626.5** MOBILE-SATELLITE (Earth-to-space) AERONAUTICAL RADIONAVIGATION RADIODETERMINATION- SATELLITE (Earth-to-space) Mobile-Satellite (space-to-Earth) 722 731E 731F 732 733 733A 733C 733D 733E	**1 613.8 – 1 626.5** MOBILE-SATELLITE (Earth-to-space) AERONAUTICAL RADIONAVIGATION Radiodetermination-Satellite (Earth-to-space) Mobile-Satellite (space-to-Earth) 722 727 730 731E 731F 732 733 733A 733B 733E
1 626.5 – 1 631.5 MARITIME MOBILE- SATELLITE (Earth-to-space) Land Mobile-Satellite (Earth-to-space) 726B 722 726A 726D 727 730	**1 626.5 – 1 631.5** MOBILE-SATELLITE (Earth-to-space) 722 726A 726C 726D 727 730	
1 631.5 – 1 634.5	MARITIME MOBILE-SATELLITE (Earth-to-space) LAND MOBILE-SATELLITE (Earth-to-space) 722 726A 726C 726D 727 730 734A	
1 634.5 – 1 645.5	MARITIME MOBILE-SATELLITE (Earth-to-space) Land Mobile-Satellite (Earth-to-space) 726B 722 726A 726C 726D 727 730	
1 645.5 – 1 646.5	MOBILE-SATELLITE (Earth-to-space) 722 726D 734B	
1 646.5 – 1 656.5	AERONAUTICAL MOBILE-SATELLITE (R) (Earth-to-space) 722 726A 726D 727 729A 730 735	

MHz
1 656.5 – 1 675

Allocation to Services		
Region 1	Region 2	Region 3
1 656.5 – 1 660	LAND MOBILE-SATELLITE (Earth-to-space) 722 726A 726D 727 730 730A 730B 730C 734A	
1 660 – 1 660.5	LAND MOBILE-SATELLITE (Earth-to-space) RADIO ASTRONOMY 722 726A 726D 730A 730B 730C 736	
1 660.5 – 1 668.4	RADIO ASTRONOMY SPACE RESEARCH (passive) Fixed Mobile except aeronautical mobile 722 736 737 738 739	
1 668.4 – 1 670	METEOROLOGICAL AIDS FIXED MOBILE except aeronautical mobile RADIO ASTRONOMY 722 736	
1 670 – 1 675	METEOROLOGICAL AIDS FIXED METEOROLOGICAL-SATELLITE (space-to-Earth) MOBILE 740A 722	

MHz
1 675 – 1 930

Allocation to Services		
Region 1	Region 2	Region 3
1 675 – 1 690 METEOROLOGICAL AIDS FIXED METEOROLOGICAL-SATELLITE (space-to-Earth) MOBILE except aeronautical mobile 722	1 675 – 1 690 METEOROLOGICAL AIDS FIXED METEOROLOGICAL-SATELLITE (space-to-Earth) MOBILE except aeronautical mobile MOBILE-SATELLITE (Earth-to-space) 722 735A	1 675 – 1 690 METEOROLOGICAL AIDS FIXED METEOROLOGICAL-SATELLITE (space-to-Earth) MOBILE except aeronautical mobile 722
1 690 – 1 700 METEOROLOGICAL AIDS METEOROLOGICAL-SATELLITE (space-to-Earth) Fixed Mobile except aeronautical mobile 671 722 741	1 690 – 1 700 METEOROLOGICAL AIDS METEOROLOGICAL-SATELLITE (space-to-Earth) MOBILE-SATELLITE (Earth-to-space) 671 722 735A 740	1 690 – 1 700 METEOROLOGICAL AIDS METEOROLOGICAL-SATELLITE (space-to-Earth) 671 722 740 742
1 700 – 1 710 FIXED METEOROLOGICAL-SATELLITE (space-to-Earth) MOBILE except aeronautical mobile 671 722	1 700 – 1 710 FIXED METEOROLOGICAL-SATELLITE (space-to-Earth) MOBILE except aeronautical mobile MOBILE-SATELLITE (Earth-to-space) 671 722 735A	1 700 – 1 710 FIXED METEOROLOGICAL-SATELLITE (space-to-Earth) MOBILE except aeronautical mobile 671 722 743
1 710 – 1 930	FIXED MOBILE 740A 722 744 745 746 746A	

TABLE 3

Sample Page of Footntoes from Article 8 of the

International Radio Regulations

714   *Additional allocation:* in Canada and the United States, the bands 1 240 - 1 300 MHz and 1 350 - 1 370 MHz are also allocated to the aeronautical radionavigation service on a primary basis.

715   *Additional allocation:* in Indonesia, the band 1 300 - 1 350 MHz is also allocated to the fixed and mobile services on a primary basis.

716   *Alternative allocation:* in Ireland and the United Kingdom, the band 1 300 - 1 350 MHz is allocated to the radiolocation service on a primary basis.

717   The use of the bands 1 300 - 1 350 MHz, 2 700 - 2 900 MHz and 9 000 - 9 200 MHz by the aeronautical radionavigation service is restricted to ground-based radars and to associated airborne transponders which transmit only on frequencies in these bands and only when actuated by radars operating in the same band.

718   In making assignments to stations of other services, administrations are urged to take all practicable steps to protect the spectral line observations of the radio astronomy service from harmful interference in the band 1 330 - 1 400 MHz. Emissions from space or airborne stations can be particularly serious sources of interference to the radio astronomy service (see Nos. 343 and 344 and Article 36).

719   In Bulgaria, Mongolia, Poland, the German Democratic Republic,
WARC-92   Romania, Czechoslovakia and the U.S.S.R., the existing installations of the radionavigation service may continue to operate in the band 1 350 - 1 400 MHz.

720   The bands 1 370 - 1 400 MHz, 2 640 - 2 655 MHz, 4 950 - 4 990 MHz and 15.20 - 15.35 GHz are also allocated to the space research (passive) and earth exploration-satellite (passive) services on a secondary basis.

721   All emissions in the band 1 400 - 1 427 MHz are prohibited.

722   In the bands 1 400 - 1 727 MHz, 101 - 120 GHz and 197 - 220 GHz, passive research is being conducted by some countries in a programme for the search for intentional emissions of extra-terrestrial origin.

722A   Use of the band 1 452 - 1 492 MHz by the broadcasting-satellite service,
WARC-92   and by the broadcasting service, is limited to digital audio broadcasting and is subject to the provisions of Resolution 528 (WARC-92).

ATTACHMENT A

By

J. A. Klobuchar

USAF Geophysics Lab

Dear Keith,

This note is in response to your request for ways to judge what new second frequency should be used on a future GPS-type signal to provide the civilian community with ionospheric corrections. Basically, the relationship between ionospheric group delay and carrier phase advance is simply:

$$\Delta \phi = -f \Delta t \tag{1}$$

For every cycle of carrier phase advance, there are: $\frac{1}{f}$ seconds of time delay. Differential carrier phase shift at a proposed new second frequency, f_x, with the present L_1 frequency called f_1, can be expressed as:

$$\Delta \delta_\phi = \frac{1.34 \times 10^{-7}}{f_x} \times \frac{(m^2 - 1)}{m^2} \times TEC \ ... \ (cycles) \tag{2}$$

where $m = \frac{f_1}{f_x}$

Differential group delay can be expressed as:

$$\Delta \delta_{gd} = \frac{40.3}{c} \times TEC \times \left(\frac{1}{f_x^2} - \frac{1}{f_1^2} \right) \tag{3}$$

If we compute the differential phase shift and the differential group delay for different values of the secondary frequency, for a specific value of Total Electron Content, (TEC), which corresponds to one meter of range delay at L_1, ($6.15 \times 10^{16} \frac{el}{m^2}$), we can see what frequency difference you must have to be able to determine this minimum amount of range delay. I have run a simple program to compute both the differential phase shift and the differential group delay as a function of the second frequency, f_x. The attached table gives the results of those computations in the following form:

Col 1: f_x, in GHz,
Col 2: Differential carrier phase, $\Delta \delta_\phi$, in cycles at f_x,
Col 3: Differential group delay, $\Delta \delta_{gd}$ (in ns.)
A plot of those values, as a function of frequency difference, is also attached to this letter.

A differential phase change of a fraction of a cycle should not be difficult to measure, but, because of unknown differential phase changes in many portions of any such system, one cannot hope to make *absolute* differential phase measurements. Relative measurements of ionospheric changes should be readily measureable down to, say, one tenth of a cycle of differential carrier phase, so that a choice of f_x fairly near the present L_1 frequency should suffice to make measurements of only *relative* ionospheric changes. I remember from an old STEL receiver we used several years ago that the noise in the differential carrier phase measurements, using a 16 Hz bandwidth, was approximately 0.1 radians, or only 0.02 cycles, so I know that this is possible. However, the measurement of *absolute* differential group delay is another story.

Remember that, in order to see a one nanosecond differential group delay, you must be able to measure to approximately 0.01 of a cycle of a 10.23 MHz modulation envelope, or to 0.001 of a cycle, of a 1.023 MHz modulation frequency. I don't know how easy it is to measure precisely to one thousandth of a cycle of two "noisy" 1 MHz signals, but I would doubt whether this is possible, unless the signal to noise ratio is very high, which it isn't in the usual navigation system. One cannot integrate for long periods to gain signal to noise ratio when on a dynamic platform, or when the ionosphere may be changing rapidly, either. Even with the 10.23 MHz modulation it may be difficult to discern a differential of only 0.01 cycles in the presence of noise. I have neglected any multipath effects here also. You should ask someone with more receiver experience than I have whether they think that they can reliably measure *absolute* differential group delays at the sub-nanosecond level. Off hand, I would doubt that this is possible at the present time. I would recommend that a modulation frequency of 10.23 MHz be retained, and that the frequency spacing be at least 200 MHz.

I hope that these numbers assist you in the choice of a second frequency on a future commercial navigation system similar to the present GPS.

Sincerely,

Jack K.

Jack Klobuchar

COMPUTATION OF DIFFERENTIAL CARRIER PHASE ADVANCE AND
DIFFERENTIAL GROUP DELAY, VERSUS CHOICE OF SECONDARY FREQUENCY (Fx).

(J. A. Klobuchar, December 1994)

TEC = 0.6150E+17 el/m^2 column
(Equivalent to 1 meter of ionospheric range error at L1)

Carrier phase change at L1 = -5.23 cycles
Group delay at L1 = 3.33 nanoseconds

Fx (GHz)	Ddphi at FX (cycles)	Diff. G.D. (ns)	
1.56519	0.068	0.04	
1.55496	0.137	0.09	
1.54473	0.206	0.13	
1.53450	0.275	0.18	
1.52427	0.345	0.23	
1.51404	0.416	0.28	
1.50381	0.487	0.32	
1.49358	0.558	0.37	
1.48335	0.630	0.43	
1.47312	0.703	0.48	
1.46289	0.776	0.53	
1.45266	0.850	0.59	
1.44243	0.924	0.64	
1.43220	0.999	0.70	
1.42197	1.074	0.76	
1.41174	1.150	0.82	
1.40151	1.227	0.88	
1.39128	1.304	0.94	
1.38105	1.382	1.00	
1.37082	1.460	1.07	
1.36059	1.539	1.13	
1.35036	1.619	1.20	
1.34013	1.700	1.27	
1.32990	1.781	1.34	
1.31967	1.863	1.42	
1.30944	1.946	1.49	
1.29921	2.029	1.57	
1.28898	2.114	1.64	
1.27875	2.199	1.72	
1.26852	2.285	1.81	
1.25829	2.371	1.89	
1.24806	2.459	1.98	
1.23783	2.548	2.06	
1.22760	2.637	2.15	PRESENT L2 FREQUENCY
1.21737	2.727	2.25	
1.20714	2.819	2.34	
1.19691	2.911	2.44	
1.18668	3.004	2.54	
1.17645	3.099	2.64	
1.16622	3.194	2.75	
1.15599	3.291	2.85	
1.14576	3.388	2.96	
1.13553	3.487	3.08	
1.12530	3.587	3.20	
1.11507	3.688	3.32	
1.10484	3.791	3.44	
1.09461	3.894	3.57	

248

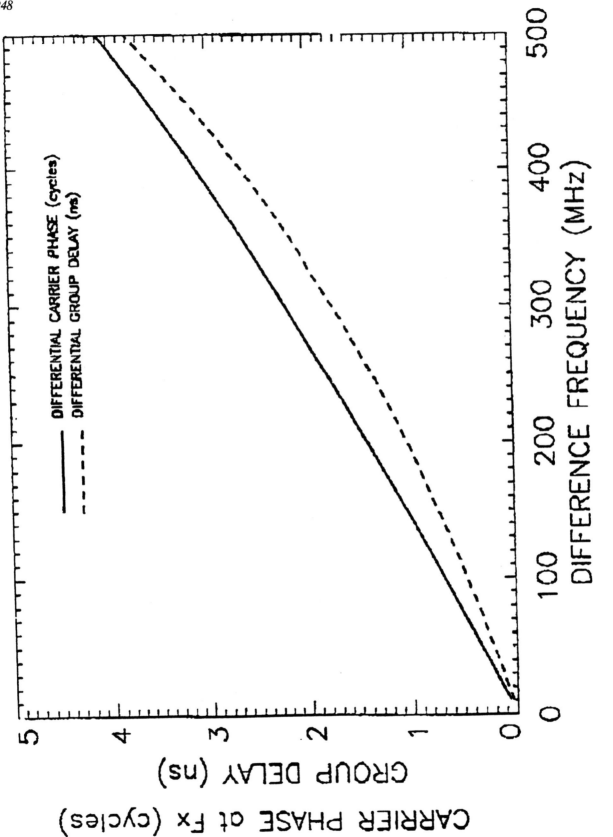

Appendix J

Selective Denial of Civilian GPS Signals by the Military

The recommended policy on GPS signal denial, in lieu of the use of SA, will force our military to take a variety of steps to deny local GPS access. Therefore, in considering signal structure enhancements of C/A-code emissions on or near L_2 or transmissions at a new frequency in the L-band, the NRC committee was mindful of the need to maintain flexible selective denial options. In all cases, the issues of whether a C/A-code signal on L_1 or L_2 could be selectively denied without severely impacting the existing military receiver inventory,, or whether a modified or enhanced military receiver would be needed were addressed. The preliminary assessment of these issues involved discussions with appropriate experts in the military GPS community (Captain Jay Purvis, National Air Intelligence Center; Mr. John Clark, the Aerospace Corporation; and Mr. William Delaney, MIT Lincoln Laboratory) as well as computer modeling of the selective denial jammer problem.

The following three questions were posed to the cognizant GPS experts:

(1) Is it feasible to employ a noise jammer that covers a 2-Mhz C/A-code bandwidth,
     - without unduly impacting local friendly Y-code users;
     - without modification to existing military receivers;
     - with modification to existing military receivers?

(2) Are there more sophisticated jamming signals that could render C/A-code receivers ineffective without unduly impacting friendly Y-code (and C/A-code) reception. For example, encrypted pseudo-noise jamming, which could be removed by friendly receivers?

(3) Are there high-confidence deceptive spoofing techniques which, over operationally useful areas, could render a sophisticated aided C/A-code receiver inoperative?

It was clear from discussions on these and related questions that the defense community is just now embarking on operationally oriented studies and activities addressing the efficacy of various denial techniques. Not surprisingly, those closer to the operational side are more doubtful than those in the development community that highly surgical jamming techniques for denial or spoofing of C/A-code reception on either the L_1 or L_2 frequency can be deployed. One of the major concerns is operational flexibility in the field, with or without modified GPS receivers.

To further assess these concerns, a measure of operational flexibility was developed for quantifying the relative impact of denial jamming on friendly and unfriendly forces. One such measure is the post-correlator signal-to-noise ratio (SNR) advantage for friendly forces. A directly related measure is the relative operating distance (ROD) to the denial jammer such that friendly and unfriendly units obtain equal signal-tracking margins:

$$ROD = \frac{\text{(enemy distance to denial jammer)}}{\text{(friendly distance to denial jammer)}_{\text{equal correlator SNR}}}$$

This function was computed assuming "friendly" Y-code receivers at L_2 (unmodified and modified variants) and "enemy" C/A-code receivers at or offset from L_2. Using frequency domain convolution techniques and two different postulated denial jammer spectra, the SNR after baseband code wipeoff was computed. For the modified friendly receiver variant, a bandstop filter was incorporated for further suppression of the selective-denial jammer.

First, the problem of signal denial precisely at L_1 or L_2 was considered. As an optimistic bound on possible performance, it was assumed for this case, an ideal high-pass filter cutting off above the first zero crossing of the C/A-code spectrum. This could be implemented by digital filtering of baseband sample prior to code correlation wipeoff. With this cutoff, a 1.5 dB loss in useful Y-code power is incurred. Following code wipeoff, the respective SNRs for the C/A-code and Y-code receivers differ by 31.5 dB. This translates into an ROD distance ratio of 37.5.

Operationally, the ROD could be exploited by a commander in several ways. Ideally the denial jammer would be so situated that the near-far or relative-distance geometry would be favorable, with the jammer located on board an aircraft at the battlefield periphery. The limiting case would be a space-born jammer, but this would require a sizable L-band antenna to meet the link budget. A more realistic scenario in a tactical situation would be a denial jammer close at hand. Suppose that jammer power has been set by a field commander to deny those C/A-code users within a radius of 10 kilometers of the jammer site. Friendly receivers equipped with comparable aiding and antenna augmentations would fail to operate within a radius of about 266 meters. This is the most optimistic scenario and requires substantial modification to Y-code receivers. Any ROD advantage would be eroded if inertial aiding and/or nulling antennas were employed by hostile units. Therefore, the NRC committee concluded that surgical jamming of the C/A-code centered at L_1 or L_2 would cause operationally unacceptable consequences for Y-code users.

Assuming heavy jamming of the existing C/A-code at L_1, with unacceptable impact on Y-code at that frequency, attention focused on a new civilian frequency, L_4. The goal was to provide a sufficiently large separation from L_1 for civilian ionospheric correction and adequate separation from L_2 to permit effective selective-denial jamming. In support of this new transmission frequency, a selective-denial jamming analysis was carried out for narrow-band C/A-like code transmissions offset from L_2, as well as a P-code like wider civilian signal offset from L_2. The first and third nulls of the L_1 Y-code, at 1237.6 MHz and 1257.6 MHz respectively, were examined.

Table J-1 summarizes the ROD ratio and SNR advantage under the noted assumptions. The first two options were covered above. Options 3 through 5 are with the new narrow-band civilian signal located at the first null. The most interesting result is for the shape II jammer spectrum. Without receiver modification the ROD distance ratio for this type of spectrum is 91.2 (39.2 dB SNR advantage). A modified receiver operating with the same jammer gives a ROD of 167. In Option 6 the narrow-band civilian signal is placed at the third zero crossing of the L_2 Y-code, and is denied with shaped noise jamming. Without receiver modification, the ROD is extremely large, and there is no difficulty in isolating the narrow-band jamming signal from L_2 Y-code.

Table J-1 Relative Operating Distances and Signal-to-Noise Advantage for Selective Denial Jamming Alternatives

Selective Denial Jamming Option	ROD (relative operating distance ratio)	dB Post-Correlator Advantage
Option 1 Narrow-band code on L_2; shape II jammer spectrum; no receiver modification	3.2	10
Option 2 Narrow-band code on L_2; shape II jammer spectrum; ideal high-pass filter in receiver	37.5	32
Option 3 Narrow-band code at first null of L_2; shape I jammer spectrum; no receiver modification	31.6	30
Option 4 Narrow-band code at first null of L_2; shape II jammer spectrum; no receiver modification	91.2	39
Option 5 Narrow-band code at first null of L_2; shape II jammer spectrum; fourth order band-stop filter	167	45
Option 6 Narrow-band code at third null of L_2; shape II jammer spectrum; no receiver modification	7,080	77
Option 7 Wide-band code at third null of L_2; shape II jammer spectrum; no receiver modification	63	36
Option 8 Wide-band code at third null of L_2; shape II jammer spectrum; receiver low pass modification	630	56

a.   Both narrow- and wide-band codes have sinc² spectrum and 1 MHz and 10 MHz chipping rates, respectively.

b.   Shape I jammer follows sinc² spectrum.

c.   Shape II jammer follows MSK (minimum shift key) spectrum.

Wide-band civilian transmissions at the third null were examined in Options 7 and 8. Neglecting receiver radio frequency/intermediate frequency selectivity, the isolation advantage from the code wipeoff process alone is 36 dB, corresponding to an ROD of 63. Typical Y-code receivers have substantial intermediate frequency attenuation 20 MHz to 30 MHz above L_2. Modeling this as an ideal low-pass starting at the 20 MHz point gives a 56 dB advantage and a corresponding ROD 630, which are adequate for any operational scenario.

The other alternatives discussed with GPS experts, but not analyzed due to time constraints, were pseudonoise jamming and spoofing. These techniques could obviously be applied in conjunction with the jamming techniques above. Pseudonoise jamming requires a modified receiver that coherently estimates and subtracts denial jamming prior to the code-correlation process. Note that this technique might offer the distinct advantage of C/A-code operation for friendly forces under certain circumstances, while denying C/A-code to the adversary. Such a technique fits well with the digital band-stop filtering incorporated in the above analysis or could be introduced at an existing receiver's radio frequency input. While it may be possible to subtract much of the pseudonoise from friendly receivers, perhaps assisted by known selective-jammer location and known user motion, experts expressed concern with null depths and the ability to rapidly adapt to multiple jammers. Once again, the sophisticated user could employ antenna nulling and receiver aiding techniques to greatly diminish the effectiveness of this kind of selective denial.

GPS signal spoofing of the so-called "denial" type, in which individual tracking loops are forced back into reacquisition mode, also was a technique discussed with the GPS experts. It was possible to postulate a number of techniques that would reduce its effectiveness; therefore, this technique, taken by itself, was not considered as adequate for selective denial.

The above techniques are illustrative of the potential denial techniques that could be applied operationally. Denial jamming of an offset L_2 frequency offers clear advantages over the other techniques. However, further in-depth study may suggest ways to combine these techniques for greater operational effectiveness and flexibility.

Appendix K

Direct Y-Code Acquisition

Below are calculations showing the time for direct Y-code acquisition with older application specific integrated circuit (ASIC) technology and current ASIC technology. In the analysis, the following assumptions were made:

(1) Receivers have limited knowledge of their current position.
(2) Receivers are using the latest satellite ephemerides.
(3) Time is known to $\pm$ 1 second.

OLD TECHNOLOGY (100,000 GATE ASIC)

The Y-code has 10^7 chips to search, given a 1-second uncertainty in clock offset (10.23 million chips per second). A well-designed receiver can obtain a signal-to-noise ratio of 12.6 dB in 0.001 seconds, based on the following derivation:

$$\text{Noise power} = kTB,$$

where k is Boltzman's constant, k = -198.6 dBm/Hz/Kelvin or $10^{-19.86}$ milliwatts/Hz/K. Assume the system temperature, T, is 100 Kelvin, then B, the noise bandwidth, is taken to be 1/0.001 seconds, or 1,000 Hz. Thus:

$$
\begin{aligned}
\text{Noise power} &= (10^{-19.86}\,\text{milliwatts/Hz/K})(1{,}000\,\text{Hz})(100\,\text{K}) \\
&= 10^{-14.86}\,\text{milliwatts or } -148.6\,\text{dBm}
\end{aligned}
$$

Given the minimum received power level for the L_2 signal, which is -136 dBm, the ratio of signal-to-noise can be calculated:

$$
\begin{aligned}
\text{Signal-to-noise} &= \text{received power} - \text{noise power} \\
&= -136\,\text{dBm} - (-148.6\,\text{dBm}) \\
&= 12.6\,\text{dB.}
\end{aligned}
$$

12.6 dB is more than adequate for detection, which means that the ratio of signal voltage-to-noise is 4.3. If the detection threshold were conservatively set at three times the noise there would only be a 1-three sigma, or about 1 percent probability of false detection.

If a receiver is implemented with a parallel search capability of 1,000 correlation channels, a full search over 1 second of delay could be accomplished in 10 seconds based on the equation below.[1]

$$(10^7 \text{ chips})(0.001 \text{ correlation channel sec/chip search})/(1,000 \text{ correlation channel})$$
$$= 10 \text{ seconds.}$$

This assumes that the signal Doppler is known to about 1,000 Hz, which corresponds to about 200 m/second, or 720 km/hr.

CURRENT TECHNOLOGY (500,000 GATE ASIC)

The search time would be reduced by a factor of 5, to 2 seconds. Using the same procedure as above, if a receiver is implemented with a parallel search capability of 5,000 correlation channels, a full search over 1 second of delay could be accomplished in

$$(10^7 \text{ chips})(0.001 \text{ correlation channels sec/chip search})/(5,000 \text{ correlation channel})$$
$$= 2 \text{ seconds.}$$

Again, this assumes that the signal Doppler is known to 1,000 Hz, which corresponds to about 200 m/second, or 720 km/hr.

DISCUSSION

For both cases, modest assumptions about receiver capabilities have been made. Time keeping accurate to 1 second is within the range of a wristwatch-level oscillator over a day or so. Most platforms can estimate their velocity to 720 km/hr. If the velocity and time are not known to this level, additional multiples of the 10- or 2-second search would be required. Once the first satellite is acquired, the receiver clock can be fixed to about 0.01 second, so searches for additional satellites can be done sequentially taking about 0.1 second each. We have also assumed that the receiver has on-board ephemerides for the satellites to allow position solutions immediately following acquisition of the first four satellites. If there are no on board ephemerides, it takes about 30 seconds to receive all five ephemeris subframes, so 30 seconds should be added to obtain a time-to-first-fix.

---

[1] A chip to perform the parallel search would require about 100,000 gates if implemented in a gate array, and these have been available for many years. (For comparison, 500,000 gate arrays are now available.) About 50,000 gates would be required to implement 1,000 correlation channels in a more efficient full-custom ASIC.

Appendix L

Enhanced Signal Structures for the Military

A significant increase (approximately 10 dB) in anti-jam capability could possibly be achieved on the Block IIF satellites by employing another wide-band signal, occupying perhaps 100 MHz to 200 MHz. Such a broad signal would require that the carrier be at S-band (approximately 3 GHz) or higher frequency. The move to a higher frequency also would reduce nulling antenna size and increase its performance. Such a high frequency would also provide increased immunity to the effects of ionospheric scintillation, which can degrade receiver performance when it is present.[1]

To demonstrate the anti-jam effectiveness of a wide-band, fine ranging signal, calculations for seven possible signal scenarios (with various bandwidths, antennas, and inertial aiding) have been performed for jammers operating at power levels of 100 watts and 10 kilowatts. In each case, the jammers were assumed to be co-located with the target. At these two power levels, code- and carrier-tracking thresholds were estimated as a function of range from the jammer. For many applications, the key parameter is not the minimum range for signal lock, but the minimum range for acceptable range error. Therefore, the minimum range-to-jammer for a 1-meter range error was also determined. It is important to distinguish two quite different operating scenarios: direct attack and loiter. In direct attack, the range-to-target is closed as rapidly as possible. Once GPS is lost, guidance to the target is by inertial guidance alone. Mission success then depends upon the remaining distance to target as well as the inertial drift rate. By contrast, in loitering scenarios such as remotely piloted vehicle reconnaissance and other scenarios involving sustained area-wide high accuracy, loss of GPS means loss of high accuracy positioning, as inertial drifts can quickly exceed mission error bounds.

Table L-1 summaries the seven signal scenarios. Scenario 1, 2, and 3 with Y-code signaling (20-MHz bandwidth) were considered as baseline for comparison with the other scenarios, each with a 100-MHz chipping rate (200-MHz bandwidth). A high chipping rate direct-sequence modulation was chosen to improve both the jamming margin and pseudorange accuracy. Under the assumption that a wide region of the L-band would be hard to come by and that beam-forming antenna structures are large at L-band, a fourfold

---

[1] Ionospheric scintillation is a phenomenon in which the Earth's ionosphere introduces rapid phase and amplitude fluctuations in the received signals.

frequency increase was predicted. In each scenario, attention was given to the thermal noise limited region and the interference limited region. For military users in a combat environment, receiver and thermal noise is negligible compared with jamming power.

Table L-1 Summary of Seven Signal Scenarios with Different Bandwidths, Antennas, and Inertial Aiding

Scenario	Bandwidth	Antenna Used	Inertial Aiding	Code Loop Tracking Bandwidth	Carrier Loop Tracking Bandwidth
1 (Baseline)	Y-code Bandwidth (20 MHz)	Standard Antenna	No	1.0 Hz	20 Hz
2 (Baseline)	Y-code Bandwidth (20 MHz)	Standard Antenna	Yes	1.0 Hz (aided)	1.0 Hz (aided)
3 (Baseline)	Y-code Bandwidth (20 MHz)	Nulling Antenna (25 dB nulls)	Yes	0.1 Hz (aided)	1.0 Hz (aided)
4	Wide Bandwidth (200 MHz)	Standard Antenna	No	1.0 Hz	20 Hz
5	Wide Bandwidth (200 MHz)	Standard Antenna	Yes	1.0 Hz (aided)	1.0 Hz (aided)
6	Wide Bandwidth (200 MHz)	Miniature Antenna (25 dB nulls)	Yes	0.1 Hz (aided)	1.0 Hz (aided)
7	Wide Bandwidth (200 MHz)	Null/ Beamforming Antenna (31 dB nulls and 6 dB beam gain)	Yes	0.1 Hz (aided)	1.0 Hz (aided)

SCENARIO 1: UNAIDED Y-CODE BANDWIDTH SIGNAL WITH A STANDARD ANTENNA

For comparison purposes, a baseline of an unaided Y-code bandwidth GPS receiver operating with a standard antenna will be used.

SCENARIO 2: AIDED Y-CODE BANDWIDTH WITH A STANDARD ANTENNA

For comparison purposes, a baseline of an aided Y-code bandwidth GPS receiver operating with a standard antenna will be used.

SCENARIO 3: AIDED Y-CODE BANDWIDTH WITH NULLING ANTENNA

For comparison purposes, a baseline of an aided Y-code bandwidth GPS receiver operating with a nulling antenna will be used.

SCENARIO 4: UNAIDED WIDE BANDWIDTH WITH STANDARD ANTENNA

This scenario is compared with the baseline described in Scenario 1.

Receiver Thermal Noise Limited Case

In this condition, the four times higher radio carrier frequency will give a free-space carrier-to-noise ratio disadvantage of 12 dB. Above the code-tracking loop threshold, the 12 dB loss is more than offset by increased signal bandwidth. Multipath susceptibility is reduced by factors of 10 and 100, respectively, over Y-code and C/A-code.

Noise Jammer Limited Case

Importantly, any increase in free-space loss with frequency is equal for both the interference source and the GPS satellite. With the narrower code chip of the wide-band signal structure, better calibration of the constellation will be needed.

SCENARIO 5: AIDED WIDE BANDWIDTH STANDARD ANTENNA

The comparative baseline is the aided Y-code receiver operating with a standard antenna, Scenario 2.

Receiver Thermal Noise Limited Case

In this condition, the four times higher radio carrier frequency will give a free-space carrier-to-noise ratio disadvantage of 12 dB. Above the code-tracking loop threshold, the 12 dB loss is more than offset by increased signal bandwidth. Multipath susceptibility is reduced by factors of 10 and 100, respectively, over Y-code and C/A-code.

Noise Jammer Limited Case

As for Scenario 4, any increase in free-space loss with frequency is equal for both the interference source and the GPS satellite. Therefore the wider bandwidth yields a 10-dB advantage in break-lock margin and a further operational advantage at a specified pseudorange accuracy level.

SCENARIO 6: AIDED WIDE BANDWIDTH WITH MINIATURIZED NULLING ANTENNA

The comparative baseline is the aided Y-code receiver operating its nulling antenna, Scenario 3. For equal nulling performance, a fourfold increase in radio frequency would reduce the overall antenna footprint to one-sixteenth the original area, making for a much more practical design in many applications. With aiding, the code- and carrier-tracking loop bandwidths are conservatively reduced to 0.1 Hz and 1 Hz, respectively.

Receiver Thermal Noise Limited Case

Same comments as for Scenario 4.

Noise Jammer Limited Case

As in Scenario 5, the widened signal bandwidth gives an immediate improvement in effective carrier-to-noise ratio of 10 dB against the reference system and a consequent 10-dB increase in jamming-to-signal ratio code and carrier tracking margin. As shown in Table L-1, this factor, together with narrowed tracking loop bandwidths yields a factor of three improvement in minimum jammer distance before loss of lock. More importantly, a factor of six reduction in jammer distance to the 1-meter error threshold is obtained. These results are achieved with a much smaller antenna than at L_1.

SCENARIO 7: AIDED WIDE BANDWIDTH WITH NULLING AND BEAM-FORMING ANTENNA

The baseline is Scenario 3, an aided Y-code receiver operating with a null-steering antenna. The size advantages of Scenario 4 are now given up in favor of a wide-band antenna possessing four times as many elements. This translates into more (and deeper) nulls and the capacity to form beams in the direction of GPS satellites. It is assumed that nulls are improved by 6 dB over the reference antenna and that a 6-dB gain may be obtained in the direction of each satellite. Obviously these parameters need future study and verification.

Receiver Thermal Noise Limited Case

Because of antenna beam-forming, there is just a 6-dB loss in carrier-to-noise ratio as compared with the reference Y-code system. Above tracking threshold this loss is more than offset by increased signal bandwidth, with an order of magnitude ranging error improvement.

Noise Jammer Limited Case

This is the most important case. Over the reference system, the widened signal bandwidth gives an immediate improvement in effective carrier-to-noise ratio of 10 dB. To this add 12 dB from improved antenna nulling and beamforming, for a total of 22 dB increase in the jamming-to-signal ratio code- and carrier-tracking margin. As shown in Tables L-2 and L-3, and Figures L-1 and L-2, there is an order of magnitude improvement in minimum jamming distance before loss of lock and a factor of 20 improvement in minimum jamming distance at 1-meter error threshold.

Figures L-1 and L-2 show the pseudorange errors as a function of distance for various receiver alternatives described in Table L-1 and the two jammer power levels.[2] The difference between the Y-code and wide-band options is rather dramatic, even on the log-log plots. The most capable system operates below the 1-meter level to within about 45 meters of the 100-watt source. At 1,000 meters, the code-tracking error is below the centimeter level. As shown in Table L-2, carrier-phase tracking and code-loop aiding are available within several hundred meters of the jammer. The miniaturized nulling antenna with aiding is good down to about 175 meters. Both aided wide-band options are substantially more capable than the best performing existing Y-code system.

Tables L-2 and L-3 summarize the results of this exercise. The most significant finding, perhaps, is that with the wide-band signal using unaided tracking and a simple antenna a vehicle can approach a 100-watt jammer to within 6 kilometers before a 1-meter range error has accumulated. With aided tracking, this range is reduced to about 3 kilometers. For many airborne weapons systems this is sufficiently close to permit a successful mission when employing inertial navigation for the balance of the flight (i.e., assuming the worst case scenario in which the jammer and target are co-located). Considering that the size and cost of nulling antennas may prohibit their use on certain weapon systems, this is a significant finding and supports the notion that consideration should be given to the eventual inclusion of a new, very wide-band waveform. Note also that a move to higher frequency makes the nulling antenna more feasible for many vehicles. As a means of defeating enemy jamming, the Air Force should explore the feasibility of adding

---

[2] Data generated by J. W. Sennott, Bradley University, Peoria, Illinois.

a new wide-band ranging signal on Block IIF satellites operating at S-band or higher frequency.

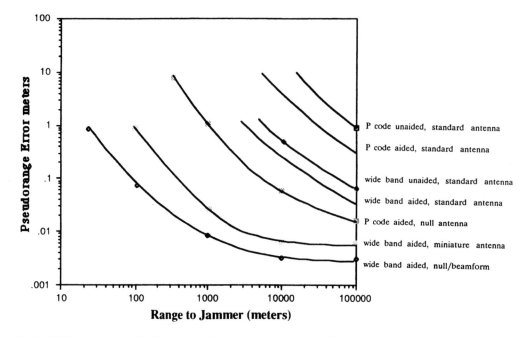

Figure L-1 Wide-band GPS with 100-watt jammer.

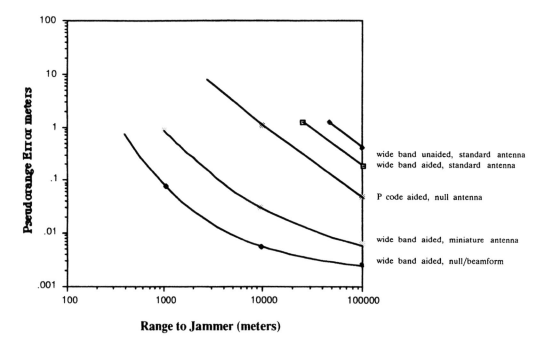

Figure L-2 Wide-band GPS with 10-kilowatt jammer.

Table L-2 GPS Wide-Band Signal Augmentation Performance 100-Watt Jammer

System Option	Code Status		Carrier Telemetry Status	
	Jammer distance at loss of lock (meters)	**Jammer distance for 1-meter range error (meters)**	**Jammer distance at loss of lock (meters)**	**Range error at loss of lock (meters)**
1. Y-code unaided standard antenna	18,000	90,000	90,000	1.0
2. Y-code aided standard antenna	10,000	35,000	21,000	---
3. Y-code aided nulling antenna	550	1,000	1,400	1.9
4. Wide-band unaided standard antenna	6,000	6,000	35,000	0.1
5. Wide-band aided standard antenna	3,100	3,100	6,500	0.27
6. Wide-band aided miniature antenna	175	175	450	0.19
7. Wide-band aided null/beamforming antenna	45	45	215	0.19

Table L-3 GPS Wide-Band Signal Augmentation Performance 10-Kilowatt Jammer

System Scenario	Code Status		Carrier Telemetry Status	
	Jammer distance at loss of lock (meters)	Jammer distance for 1-meter range error (meters)	Jammer distance at loss of lock (meters)	Range error at loss of lock (meters)
1. Y-code unaided standard antenna	---	---	---	---
2. Y-code aided standard antenna	---	---	---	---
3. Y-code aided nulling antenna	---	20,000	---	---
4. Wide-band unaided standard antenna	---	60,000	---	---
5. Wide-band aided standard antenna	---	31,000	---	---
6. Wide-band aided miniature antenna	---	1,800	---	---
7. Wide-band aided null/beamforming antenna	---	450	---	---

Appendix M

Accuracy of a 14-Satellite Ensemble
Versus a 24-Satellite Ensemble

Below is a comparison of the accuracy of a 14-satellite ensemble clock versus a 24-satellite ensemble.

CASE A

Assume that all satellites have clocks equal to Block IIR cesium clocks. (Block IIR rubidiums are a factor of two more stable.) A 14-satellite ensemble is used. Consider synchronization error between two satellites whose ensembles have the minimum overlap of four. Note that these satellites are on opposite sides of the earth, and would probably never be used in the same stand-alone solution, so this is the worst case scenario.

Analysis

For $T = 15$ minutes, $\Delta f/f = 10^{-12}$ Allan variance slope is $-1/2$. Autonomous navigation ranging error is 1 ns, measured each 15 minutes.

To determine the optimum clock averaging interval if (1) $T = 15$ minutes; (2) ranging error is $1\text{ns}/N^{1/2}$; (3) N is the number of 15 minute ranging epochs used for averaging; and (4) the error due to clock instability is $[(10^{-12})(1/N^{1/2})(N)$ intervals $\times$ 900 s/interval], the optimum is about 15 minutes, where measurement error and clock instability each contributes about 1 ns of error. The produces a combined (RSS) error of 1.4 ns or 0.4 meters.[1]

Given that a 14-satellite ensemble is quite adequate for the case in which all clocks are well-behaved atomic standards (rubidium or cesium), it seems evident that an ensemble of all the clocks is better. First, it will have marginally smaller error, by $(14/24)^{1/2} = 0.76$. Second, it will compare all satellite clocks at each autonomous navigation measurement,

---

[1] 1 nanosecond times the speed of light = 30 centimeters

giving improved potential for autonomous fault detection and system stability characteristics in the presence of anomalous behavior.

CASE B

If quartz oscillators with $\blacktriangle f/f = 10^{-11}$ are used with 900 s inter-satellite link ranging updates, a 14-satellite ensemble would allow significant differences (few ns) to exist among the ensemble clocks of different satellites. If a 14-satellite ensemble is used, consider synchronization error between two satellites whose ensembles have *no overlap*. (Only because this is easier to analyze. The real case is not this bad). Again, note that these satellites are on opposite sides of the Earth, and would probably never be used in the same stand-alone solution.

Analysis

(1) For a 14-satellite ensemble:

$$\text{error per clock } (\sim 10^{-11})(900 \text{ s})N = (9 \text{ ns})N$$

where:

> N is the number of 15-minute intervals that this minimum overlap occurs. For a 4-hour period, $N = 16$.

> When averaged over 14 clocks, the error would be reduced to:
> $(9\text{ns})(16/14^{1/2}) = 38$ ns.

Also, the 38 ns would not only show up as an offset from UTC, but would add to the UERE and, thus, affect the stand-alone position solution. Although as mentioned above, the real case would not be this bad.

(2) For a full constellation 24-satellite ensemble:

The clock error of the full constellation would drift by $[\{(10^{-11})(3600)(4)\}/24^{-1/2} = 29$ ns] over the same 4-hour period. While this 29 ns drift would show up as an offset from UTC, it would be a common clock error for the entire constellation, and would not significantly affect the stand-alone position solution.

In summary, the main reason for a 24-satellite clock ensemble is to enable use of more reliable, lower mass and power quartz oscillators in most of the satellites. Atomic clocks would be used in four satellites to provide redundant steering of the ensemble to UTC.